SOCIAL INNOVATION
AND TERRITORIAL DEVELOPMENT

To Len Arthur,
a great friend and the radical uncle we never had.

Social Innovation
and Territorial Development

Edited by

DIANA MacCALLUM
Griffith University, Australia

FRANK MOULAERT
Katholieke Universiteit Leuven, Belgium

JEAN HILLIER
University of Newcastle upon Tyne, UK

SERENA VICARI HADDOCK
University of Milan – Bicocca, Italy

ASHGATE

Published by
Ashgate Publishing Limited
Wey Court East
Union Road
Farnham
Surrey, GU9 7PT
England

Ashgate Publishing Company
Suite 420
101 Cherry Street
Burlington
VT 05401-4405
USA

www.ashgate.com

British Library Cataloguing in Publication Data
Social innovation and territorial development
    1. Economic development - Sociological aspects 2. Social
    change 3. Regional planning 4. Urban renewal
    I. MacCallum, Diana
    306.3

Library of Congress Cataloging-in-Publication Data
Social innovation and territorial development / edited by Diana MacCallum ...
[et al.].
    p. cm.
    Includes bibliographical references and index.
    ISBN 978-0-7546-7233-3
    1. Community development. 2. Social capital (Sociology) I. MacCallum, Diana.

    HN49.C6S633 2008
    307.1'4--dc22

                              2008039454
ISBN: 978 0 7546 7233 3

Mixed Sources
Product group from well-managed
forests and other controlled sources
www.fsc.org Cert no. SGS-COC-2482
© 1996 Forest Stewardship Council
FSC

Printed and bound in Great Britain by
TJ International Ltd, Padstow, Cornwall

Contents

List of Figures   *vii*
List of Tables   *ix*
Notes on Contributors   *xi*
Preface   *xv*

Introduction   1
Diana MacCallum, Frank Moulaert, Jean Hillier and Serena Vicari Haddock

**PART I – SOCIAL INNOVATION: NEEDS SATISFACTION,
COMMUNITY EMPOWERMENT AND GOVERNANCE**

1   Social Innovation: Institutionally Embedded, Territorially
     (Re)Produced   11
     Frank Moulaert

2   Social Innovation for Community Economies   25
     J.K. Gibson-Graham and Gerda Roelvink

3   Microfinance, Capital for Innovation   39
     Mariana Antohi

4   Civil Society, Governmentality and the Contradictions of
     Governance-beyond-the-State: The Janus-face of Social Innovation   63
     Erik Swyngedouw

PART II – CITIES AND SOCIALLY INNOVATIVE NEIGHBOURHOODS

5   Social Innovation for Neighbourhood Revitalization: A Case
     of Empowered Participation and Integrative Dynamics in Spain   81
     Arantxa Rodriguez

6   How Socially Innovative is Migrant Entrepreneurship?
     A Case Study of Berlin   101
     Felicitas Hillmann

7 Social Innovation, Reciprocity and the Monetarization of Territory in Informal Settlements in Latin American Cities 115
Pedro Abramo

8 Social Innovation and Governance of Scale in Austria 131
Andreas Novy, Elisabeth Hammer and Bernhard Leubolt

9 Inclusive Places, Arts and Socially Creative Milieux 149
Isabel André, Eduardo Brito Henriques and Jorge Malheiros

Index *167*

List of Figures

1.1	Social innovation and integrated area development	19
2.1	A diverse economy	28
3.1	Microfinance: Mobilisation of financial capital for the (re)construction of social capital	51
6.1	Unemployment rates of the German, foreign and Turkish population in Berlin, 1995–2005	105
6.2	Turkish, Polish and Vietnamese business registrations and cancellations in Berlin, 1992–2006	106

List of Tables

8.1	Two ideological camps and corporatist social partnership in Austria	136
9.1	Summary – developing Montemor-o-Novo through arts	157
9.2	Summary – playing *Wozzeck* in Aldoar	162
9.3	The plasticity of socially creative milieux: Montemor-o-Novo and Aldoar	164

Notes on Contributors

Pedro Abramo is Professor of Urban Economics at the Universidade Federal de Rio de Janeiro (Brazil). He coordinates a Latin American Network on urban informality (INFOSOLO) and is director of the Observatory of Urban Land Markets at the University of Rio de Janeiro. His most recent books include: *A Cidade Kaleidoscopica* (Bertrand do Brasil, 2005) *Cidade da Informalidad* (Rio de Janeiro: Sette Letras, 2003). He has published extensively on urban economics, urban informality and local development.

Isabel André is Professor of Human Geography at the University of Lisbon, Portugal. She coordinates a research project – LINKS – on social capital and innovation networks promoting local development at CEG-UL (Centro de Estudos Geográficos – University of Lisbon). Recently she published 'Redes y desarrollo local: la importancia del capital social y de la innovación', *Boletín de la AGE*, no. 36, 2003, pp. 117–27 (with Patricia Rego) and 'Portugal: knowledge-intensive services and modernization' in Peter Wood (ed.) *Consultancy and Innovation: the business service revolution in Europe* (London and New York: Routledge, 2002), pp. 283–315 (in collaboration with João Ferrão and Paulo Areosa Feio).

Mariana Antohi is currently finalizing her PhD in Industrial Economics at the University of Lille 1 (USTL), France. She is a member of CLERSE (a research unit of USTL and CNRS). Her research interests focus on ethics and economics, microfinance and local development, responsible economy initiatives, social inclusion/cohesion processes and the EU social policy. She has been the secretary general of the European Inter-network of Ethical and Responsible Economy Initiatives (IRIS) created within the European Dialogue Platform on ethical and solidarity-based initiatives for combating poverty and social exclusion of the Council of Europe (COE) and with the support of the European Commission.

Katherine Gibson is Professor and Head of the Department of Human Geography, Research School of Pacific and Asian Studies at the Australian National University. Under the pen name J.K. Gibson-Graham she co-authored with Julie Graham *The End of Capitalism (as we knew it)* (Blackwell, 1996; Minnesota, 2006) and *A Post-Capitalist Politics* (Minnesota, 2006). With economists Stephen Resnick and Richard Wolff JKGG co-edited *Class and Its Others* (Minnesota, 2000) and *Re/Presenting Class* (Duke, 2001). They have pursued local action research internationally and published widely on building alternative economies in place.

Julie Graham is Professor and Associate Department Head in the Department of Geosciences, University of Massachusetts Amherst. Under the pen name J.K. Gibson-Graham she co-authored with Katherine Gibson *The End of Capitalism (as we knew it)* (Blackwell, 1996; Minnesota, 2006) and *A Post-Capitalist Politics* (Minnesota, 2006). With economists Stephen Resnick and Richard Wolff JKGG co-edited *Class and Its Others* (Minnesota, 2000) and *Re/Presenting Class* (Duke, 2001). They have pursued local action research internationally and published widely on building alternative economies in place.

Elisabeth Hammer lectures at FH Campus Wien, University of Applied Sciences, Department of Social Work, with additional research experience at the Vienna University of Economics and Business Administration, Department of City and Regional Development. Her research interests focus on social policy, social work and administration, urban studies and local development.

Eduardo Brito Henriques is Professor at the Faculty of Arts, University of Lisbon and Researcher at the Centre for Geographical Studies, University of Lisbon. For over a decade, he has divided his work between urban studies, cultural geography, and leisure and tourism studies. He has published over three dozen titles on these subjects. He has also been closely involved in territorial planning as a past member of the Portuguese National Programme for Spatial Planning Policy Committee and the study group for the Re-qualification and Revitalisation of Historical Urban Centres, both under the Portuguese Department of Territorial Planning and Urban Development.

Jean Hillier is Professor of Town and Country Planning at the University of Newcastle. Her main research interests lie in praxis-based planning theories and in discursive and relational analyses of participatory planning strategies. Underlying most of her research is a deep concern with issues of social exclusion, the impact of planning decisions on women and marginalized groups. She is Managing Editor of the international journal *Planning Theory*. Publications include *Stretching Beyond the Horizon: a multiplanar theory of spatial planning and governance* (Ashgate, 2007), *Shadows of Power: an allegory of prudence in land use planning* (Routledge, 2002), and *Habitus: a sense of place* (edited with Emma Rooksby; 2nd edition, Ashgate, 2005).

Felicitas Hillmann is Professor for Human Geography at the University of Bremen, Department for Social Sciences. She has done research for many years in the field of international migration and labour market integration. Recent publications include: *Asian Migrants and European Labour Markets* (with E. Spaan and Ton van Naerssen; Routledge, 2005); 'Migrants Care Work in Private Households or: the strength of bilocal and transnational ties as a last(ing) resource in global migration' in Birgit Pfau-Effinger and Birgit Geissler (eds) *Care Work in Europe* (Blackwell, 2005). She has published widely on the mobility of the highly qualified, gender and migration, ethnic economies and labour market insertion of the immigrant population.

Bernhard Leubolt is a PhD student in Political Science at the University of Kassel, Germany in the doctoral programme Global Social Policies and Governance, with further research experience at the Vienna University of Economics and Business Administration, Department of City and Regional Development in the EU-funded project KATARSIS. Bernhard's research interests include development studies, participation, governance, state theory and inequality. His recent book is: *Staat als Gemeinwesen. Das Partizipative Budget in Rio Grande do Sul und Porto Alegre* (Wien/Münster: LIT, 2006). He has published on multi-scalar governance, participatory governance, social policies and development-related issues in Europe and Brazil.

Diana MacCallum is a Lecturer in Planning at the School of Environment, Griffith University (Queensland), and was previously a research associate for the EU-funded project *KATARSIS*, a coordination action for research on growing inequality and social innovation in Europe. Her research interests include governance practices, participation and politics, with particular emphasis on the relationship between discourse, power and action.

Jorge Malheiros is Professor of Human Geography at the Faculty of Arts, University of Lisbon. His main research interests are international migration and urban social geography. He is co-coordinator of Cluster B5 (*Social Integration and Mobility*) of the European Network IMISCOE and the Portuguese Correspondent of SOPEMI-OECD (since 2001). His recent publications include: *Espaços e Expressões de Conflito e Tensão entre Autóctones, Minorias Migrantes e Não migrantes na Área Metropolitana de Lisboa* (with Manuela Mendes; ACIDI, 2007) and 'Immigration and City Change: the region of Lisbon in the turn of the 20th century', *Journal of Ethnic and Migration Studies*, vol. 30, no. 6, November 2004, pp. 1065–86 (with Francisco Vala Salvador).

Frank Moulaert is Professor of Spatial Planning at the Katholieke Universiteit Leuven, Belgium and a Visiting Professor at the University of Newcastle upon Tyne, England. He coordinates a graduate research team at IFRESI-CNRS, Lille, France. His two most recent books are: *Integrated Area Development in European Cities* (Oxford University Press, 2000; 2002); and *The Global City: urban restructuring and social polarization in the city* (edited with Erik Swyngedouw and Arantxa Rodriguez; Oxford University Press, 2003). He has published widely on the social region and social innovation in local communities.

Andreas Novy is a Professor at the Institute of Environmental and Regional Development, Vienna University of Economics and Business Administration. His has published extensively on urban development and Latin America, e.g.

Brasilien: Die Unordnung der Peripherie (Vienna, 2001, also published in Brazil) and *Entwicklung gestalten* (Vienna, 2002, also published in El Salvador). He has participated in several FP-projects and is involved in transdisciplinary research and education, for example, as the head of the Paulo Freire Centre in Vienna.

Arantxa Rodriguez is Professor of Urban and Regional Economics and Planning at the Faculty of Economics at the University of the Basque Country, Bilbao, Spain. Her research focuses on socioeconomic and territorial restructuring and policies. Her recent publications include: *The Global City: urban restructuring and social polarization in the city* (edited with Frank Moulaert and Erik Swyngedouw; Oxford University Press, 2003) and *Territorios Inteligentes: Dimensiones y Experiencias Internacionales* (edited with M. Esteban, I. Uhalde and A. Altuzarra; Madrid: Netbiblio, 2008). She has worked extensively on local development planning and urban revitalization.

Gerda Roelvink is currently finishing her PhD in Human Geography in the Research School of Pacific and Asian Studies at The Australian National University. Her research interests focus on collective struggle and economic experimentation. She has published articles on liberalism and social policy, political economic representation and documentary films, and markets and theories of performativity.

Erik Swyngedouw is Professor of Geography at the University of Manchester. His most recent books include: *Social Power and the Urbanisation of Water* (Oxford University Press, 2004), *In the Nature of Cities* (edited with N. Heynen and M. Kaika; Routledge, 2005) and *The Global City: urban restructuring and social polarization in the city* (edited with F. Moulaert and Arantxa Rodriguez; Oxford University Press, 2003). He has published widely on political economy and spatial restructuring, urban governance and change, and on political ecology.

Serena Vicari Haddock is Associate Professor of Urban Sociology in the Department of Sociology and Social Research of the University of Milan – Bicocca. Her primary research interest is urban redevelopment and regeneration initiatives in Italian cities and in comparative perspective. Her most recent book is *La città contemporanea* (Bologna: il Mulino, 2004).

Preface

This book has come a long way. Its starting point was a workshop on social innovation, territorial and community development. The workshop was organized by IFRESI-CNRS (Lille, France) and GURU (School of Architecture, Planning and Landscape, Newcastle University), coordinator and partner, respectively, in the European Framework Programme 5, 'Social innovation, governance and community building (SINGOCOM)'. IFRESI's participation in SINGOCOM was co-funded by CNRS and the Région Nord-Pas de Calais under the Contrat Plan Etat-Région, IFRESI-CNRS. This co-funding involved a more detailed study of Alentour, a Roubaix-based association specializing in the organization of networks of social services, as well as the coordination of a workshop on the role of social innovation in neighbourhood development.

The workshop organizers wished to situate the results of SINGOCOM in a broader context and in particular to address questions about the meaning of social innovation in policy and collective action spheres other than neighbourhood and community development. To this purpose they invited speakers who focused on the cultural and political dimensions of social innovation as well as the relations between social innovation and culture as a motor of human progress and emancipation. They also confronted the SINGOCOM approach to social innovation at the neighbourhood level with other approaches, including the ABCD approach applied in the United States and Australia and the more mainstream territorial development approaches put forward by the EU and OECD.

There was, however, still some distance to cover before the papers presented at the workshop became the book manuscript. It was necessary to develop a language common to the various authors and to introduce a broader geographical perspective. This required cross-refereeing among authors as well as a confrontation with interesting cases not covered at the workshop. These exchanges resulted in a book presenting a coherent, yet pluralistic, comparison between approaches to socially innovative initiatives in neighbourhood development to date. We hope you will enjoy reading it and look forward to your feedback.

Before diving into the 'serious work' we would like to thank the organizers of the workshop in Lille and especially Prof. Gérard Gayot, Director of IFRESI, Dr Jacques Nussbaumer, Dr Oana Ailenei and Dr Bénedicte Lefèbvre as well as Mrs Fariza Marécaille who helped with the logistics of the seminar. We also would like to thank Mrs Bernadette Williams who copy-edited the manuscript.

This research was funded by the EC FP 5 under grant HPSE-CT2001-00070 and the Contrat Plan Etat Région Nord-Pas de Calais. Frank Moulaert's contribution also benefited from a Leverhulme Fellowship 2006–2007.

Introduction

Diana MacCallum, Frank Moulaert, Jean Hillier
and Serena Vicari Haddock

Over the last 30 years, and especially since the first energy crisis of the 1970s and the neoliberal *réveil*, regional development and urban regeneration strategies have traditionally swung between two poles. On the one hand (and predominantly) such strategies centred on the market–economic: economic strategies to make regions and cities more market-feasible and competitive; physical renewal strategies (large-scale redevelopment projects, new technological systems, flagship cultural districts) to strengthen the economic basis of territories; and institutional transformations (deregulation of markets and state practices, privatization of public domain and actions, promotion of public–private partnerships) to encourage capital investment in new fields of economic activity and across newly opened borders. But on the other hand, initiatives taken by local bottom-up movements in neighbourhoods exposed to economic restructuring dynamics have benefited from special socially oriented area-based collective action and programmes. These have been led by various types of public actors (civil society and state) and have often followed an 'integrated' logic of development, trying to address a diverse range of problems and needs through coordinated action. More recently, and especially under conservative governments in the United States and many European countries, neoliberal thought, discourse and practice have produced a backlash against bottom-up social development, particularly in urban neighbourhoods that have been exposed to socioeconomic restructuring.

The discourses surrounding the dominant neoliberal practices tend to represent 'the economy' as a monolithic entity, outside the realm and control of ordinary social life. However, as in this book, the economy may be regarded differently, as a set of particular social practices constituted in social institutions and activities ranging from the everyday of family and community life to norms of consumption and social regulation of work habits and professional practice. According to this view, we cannot address needs for development or change either by looking solely at economic institutions and practice or by interpreting phenomena according to a narrow economic allocation rationale. We require an alternative analysis, one which takes account of different kinds of social relations and practices from the micro to the macro level and which draws on a positive politics of empowerment, thus also shedding a new light on the different meanings and visions of the economy.

Such an analysis can be developed around the concept of *social innovation*. This concept rejects the traditional, technology-focused application of the term 'innovation', which has been central to recent European development policy, in favour of a more nuanced reading which valorizes the knowledge and cultural assets of communities and which foregrounds the creative reconfiguration of social relations.

The introduction of the social to innovation – as well as of innovation to the social – is not mere sloganeering: it is a powerful theme with deep roots in the social science and economic literatures, as well as in the socio-political practice of the previous century and the last few decades in particular, as Chapter 1 of this book elaborates. Today social innovation is an anchor concept for research in creative arts, human organization, economic diversity, neighbourhood regeneration, regional renaissance, governance and other areas. Despite the development of the concept in such a range of disciplinary perspectives, it retains at its core a key commonality: social innovation is innovation in social relations, as well as in meeting human needs. As the chapters in this volume show, such innovations can be situated at the very micro or macro level of society; they can be agenda-driven, process-determined or a mixture of both; and they often occur at the intersection of spontaneous and rationally organized movements.

About the Chapters

This volume presents a multidisciplinary set of analyses, narratives and case studies drawn from across the globe, discussing issues of social innovation in different contexts and how it affects life, societal organization and economic relations within communities, especially but not exclusively at the local and regional levels. Authors address key questions about the nature of social innovation as a process and a strategy; about what opportunities exist and may be generated for social innovation to nourish human development.

Social Innovation: Needs Satisfaction, Community Empowerment and Governance

In Part I, four chapters place social innovation in relation to the broader social science literature and to larger issues of ethics, social justice and socio-political governance. Frank Moulaert begins this project with a detailed overview of the concept in theory, empirical research and practice across various disciplines. He explains the roots of the concept and to this purpose returns to the 'Godfathers' of sociology, Max Weber and Emile Durkheim, who examined transformation in society by addressing the interplay of the economic, cultural and ethical dimensions that determine it. This 'macro-social' perspective on social innovation was further developed in the 1930s by Joseph Schumpeter in his work on social transformation and on the relationship between different readings of 'development' and the sociological frameworks in which these readings were formed. Schumpeter linked his macro-social perspective to creative micro-behaviour in different spheres of society (economy, culture and arts …). But a truly micro-perspective to social innovation had been launched much earlier by Benjamin Franklin in his search for practical solutions to the provision of services to citizens in his eighteenth-century Philadelphia. Micro and macro, process and agency, cultural and organizational perspectives have been combined both in the social change movements of the 1960s and in various neighbourhood development initiatives in Europe starting with the 'Neighbourhood in Crisis' programmes in the late 1980s. These initiatives have been an important field of observation for Moulaert and his colleagues working on the role of local development to combat poverty (under the Poverty III programme of the EC), allowing them to analyse relationships between different dimensions of development and the role of socially innovative agencies and processes as they are territorially embedded and reproduced. Moulaert explains how different approaches to territorially embedded social innovation have stressed particular tensions between human needs satisfaction, empowerment of groups of citizens and reproduction of community social relations, illustrating this through key features of the Integrated Area Development (IAD) approach developed through European Framework Programme projects URSPIC and SINGOCOM.

J.K. Gibson-Graham and Gerda Roelvink, in Chapter 2, examine the overlaps and differences between social innovation research (especially IAD) and Community Economies, an action research approach also aimed at rethinking the economic from an empowerment perspective. Community Economies is an action research approach focused on performing and instituting new forms of social and economic organization and action. It shares with the IAD approach an orientation to supporting and promoting an economic and social order that embodies contextually situated notions of social justice and democratic governance. By making this nascent and often invisible 'movement' an object of research, and by enhancing the knowledge and self-knowledge of the projects and subjects involved, both Community Economies and the Social Innovation project participate in bringing that movement into being as a transformative force worldwide.

Chapter 3 introduces another perspective. Through a discussion of microfinance practices in the developing world and, more recently, in Europe, Mariana Antohi connects the logic of social innovation with the generation and renewal of human and social capital. Given its capacity to mobilize, renew and organize various types of capital (human, social, financial), microfinance plays a dynamic role in social inclusion processes, creating a new means of interaction that, because it calls upon solidarity, reinforces social relations and cohesion. After reviewing the microfinance domain, the chapter isolates two specific types of microfinance from among the various organizational practices, focusing most particularly on solidarity-based microfinance as a social innovation. It explores a series of new challenges facing European microfinance players. In addition to some specific challenges inherent to the sector, these include a need to go beyond ideas of 'fixing' extreme poverty by identifying and experimenting with complementary approaches and synergies between microfinance and more conventional means of combating poverty and social exclusion (public policy, development NGOs, etc.), as well as new individual initiatives on the market (for example ethical saving and investment products).

To conclude the first part, Erik Swyngedouw's analysis of governance 'beyond-the-state' confronts tensions between innovative institutional arrangements and their framing within a largely market-driven political-economic context. In recent years, a proliferating body of scholarship has attempted to theorize and substantiate empirically the emergence of new formal or informal institutional arrangements that engage in the act of governing outside and beyond the state. While much of this analysis of a changing, if not new, governmentality starts from the vantage point of how the state is reorganized in response to changing socio-economic and cultural conditions and social demands for enlarged public participation, this chapter seeks to assess the consolidation of new forms of innovative governance capacity and the associated changes in governmentality in the context of the rekindling of the governance–civil society articulation that is invariably associated with the rise of a neoliberal governmental rationality and the transformation of the technologies of government. The transformation of the state/civil society relationship is situated within an analysis of consolidating neoliberal capitalism. Swyngedouw teases out the contradictory ways in which new arrangements of governance have created innovative institutions and empowered new actors, while disempowering others. Thus, these innovative arrangements of governance-beyond-the-state are fundamentally Janus-faced. Truly *socially* innovative governance, in contrast, should be backed by socio-political movements, demanding radical change and human emancipation agendas.

Cities and Neighbourhoods: Socially Innovative Territories

Part II focuses more closely on specific examples of social innovation in cities and neighbourhoods. These examples show the interplay between different dimensions of social innovation, and identify tensions and contradictions as well as instances of complementariness between institutional transformations and socially innovative practices. In Chapter 5, Aranxta Rodriguez analyses how in Spain the urban neighbourhood has re-emerged as a relevant spatial scale for analysis and policy making. She explains how over the last decade socially innovative approaches have been on the rise and fostered more integrative approaches to spatial development, stressing empowered participation among urban communities as well as social priorities in their development agendas. She discusses the case of Trinitat Nova in the periphery of Barcelona as an example of radical social innovation and bottom-up empowerment.

Next, Felicitas Hillman links the phenomenon of ethnic entrepreneurship with the concept of social innovation, showing how practices of simple survival for migrants in Berlin gradually gained formal recognition and became part of the local development strategies of the respective city administrations. She first develops a conceptualization of ethnic entrepreneurship from the existing body of scientific literature and then presents the case study of Turkish and Vietnamese migrant entrepreneurs in Berlin. The organized marketization of ethnicized events like parades and festivals is interpreted as a part of the more structural integration of 'ethnic'-strategies, originally oriented to survival, into the urban setting, and serves to illustrate the meeting of 'development from below' with institutional attempts to foster development 'from above'. This is a particularly illustrative example of the potential of social innovations – initiated in response to marginalization and unsatisfied need – to achieve wider-than-anticipated benefits and to help bring about progressive institutional change.

Pedro Abramo takes us to Latin America in Chapter 7, to explore the complex relationship between the production of social innovations from proximity-based reciprocity relations, and real estate market valorization of those community relations in poor urban areas. A 'social paradox' in this relation is identified, to the extent that social innovations which reveal the strengthening of community ties simultaneously, through the valorization of property through the real estate market in these poor areas (monetarization of the territory), can be a factor in the eventual 'expulsion' of local residents from those areas. Thus, this chapter discusses the tension between the production of social innovations based on local reciprocity networks and the weakening of those networks as a result of the expulsion of the very families who structure reciprocity relations. It introduces the critical role of the real estate market and the dynamics of residential mobility in local socio-spatial reorganization and, from there, to empowerment. The empirical focus is on *favelas* as an example of places where a lack of public action in the supply of basic goods leads to a community self-building process whose institutional framework is defined by informality, that is, the rule of informal norms in a territory lacking formal planning definition. Two dimensions of the conversion of social innovations of the informal city into elements of market valorization are discussed. The first relates to the individualization of innovations in the form of location preferences. The second is the 'locational capital' of the poor. In this case, a social innovation – the informal habitat – is converted into capital as a result of changes taking place in the intra-urban structure, which valorize new accessibility factors.

In Chapter 8, we shift from the neighbourhood to the city and national scales. Taking a historical perspective, Andreas Novy, Elisabeth Hammer and Bernhard Leubolt examine how shifting spatial arrangements of political and socioeconomic organization in Austria have created the conditions for different types of social innovation. Special emphasis is given to two emblematic moments in periods of crisis. The first period of social experimentation and empowerment was the inter-war *Red Vienna*, an alternative local state project which experimented with the empowerment of workers and diverged from the liberal–conservative national mode of development in Austria. The second moment was the period of social experimentations fuelled by an alternative movement via 'development from below' in the 1970s and 1980s. In both cases social innovation went hand in hand with a new scalar arrangement of socioeconomic organization: in the 1920s, it was a class-based local state project in a city; in the 1970s and 1980s, it was an ensemble of regional initiatives all over the country, but this time also with strong roots in peripheral areas. Both began from a bottom-linked approach to empower deprived populations, be it via class politics or social activation. In this way, the analysis in this chapter holds a structural critique of today's 'beyond-the-state' governance dynamics referred to in Chapter 4: single-issue, middle-class based movements attempt to act as a substitute for representatively elected democratic bodies, while their progressive potential to democratically shape urban development depends on an innovative interplay of state and social movements.

In the final chapter, Chapter 9, Isabel André, Eduardo Brito Henriques and Jorge Malheiros discuss social innovation related to creativity and to creative milieux. Arts and culture are two fields intrinsically associated with new ideas and with new practices which have the capacity to challenge the established order and reshape social relations. The analysis of two case studies illustrates how arts initiatives may empower peripheral communities and help to produce social innovation. The three editions of the International Symposium of Terra(cotta) Sculpture, organized by a local artistic association of Montemor-o-Novo between 1999 and 2001, show the possibilities created through dialogue between recognized artists (in this case, sculptors) and local craftspersons. These events – integrated into a municipal strategy that privileges cultural production as a key dimension of local development – have contributed considerably to both the renewal of the urban public spaces and the promotion of collective self-esteem. The *Wozzeck* and *Demolition* Operas, produced by the Education Department of *Casa da Música* in collaboration with the Birmingham Opera Company in the context of *Porto 2001 – European Capital of Culture*, involved the significant participation of people from Aldoar, a deprived neighbourhood in Oporto. This second case shows how a stigmatized local community is capable of fighting against social exclusion processes using arts as a mode of emancipatory discourse. A comparative analysis of the two cases identifies some crucial conditions for the construction of socially creative milieux.

Social Innovation as the (Re)Making of Social Space

The chapters in this volume form a constructive response to the restrictive and restricting neoliberal economic vision of spatial, economic and social change. They detail innovative alternative imaginations and lived experiences which highlight issues of solidarity, cooperation and economic, human and cultural diversity, situated both outside the market and in combination with the market, within or outside cultural–artistic arenas, and at various spatial scales of social empowerment (neighbourhood, city, region, smaller rural localities, but always acknowledging the significant and catalytic role of national and international levels of regulation and governance). As such, they shed light on the cultural and institutional modifications necessary for social innovation to flourish, including empowering local people as subjects rather than as objects of development, fostering bottom-up governance dynamics, and refocusing macro-social and macro-economic leverages to the benefit of social and cultural development at lower spatial scales.

At the same time, these chapters embody an increasingly coherent analytical apparatus capable of analysing social innovation experiences and projects. This apparatus is built on principles of justice and equity, which are socially reproduced:

- Ethical views of 'diversity as a strength', social justice and more equality of opportunity for groups of people who, over the post-1960s history, have been earmarked as deprived, excluded or inferior because of their differences from the mainstream white, middle-class model of human emancipation. In other words: the negative ethics of difference is turned into a positive ethics of cooperation and synergy among diverse communities.
- Ethics and modes of practice are not descended from heaven, but socially reproduced in particular contexts: assets for human progress are shaped by the development path of social groups and communities, and also by ongoing negotiation and cooperation between community members.

These principles oblige us to adopt an enlarged view of development which respects a plurality of economic practices seeking to achieve more 'just' economic production and allocation systems, and which treats people as active agents of their own development, albeit within circumstances shaped by wider social forces that often work against their aspirations. This view has some important implications for social innovation analysis:

- First, a conceptualization of community development within its historical and geographical dynamics. Place and space can only be taken seriously in community development analysis if their historical making and remaking are the heart of both analysis and strategy making. As a consequence, social innovation can never be analysed as belonging only to 'its' place, the place where it was generated, but as occurring within a complex web of spatial interconnections.
- Second, a theoretically informed and empirically grounded analysis of the relationship between agency and the institutional and structural dynamics that contain both opportunities and constraints for social innovation. Social innovation can never be either 'agency only' or 'institutional transformation' only; it is always conceived and analysed as an interaction between (institutional) process dynamics and typically socially innovative agency.
- Third, ethics and modes of practice are theorized and empirically analysed as being socially reproduced in particular contexts. This means that they inherit characteristics from the development path of a specific territory, but also that they are the outcome of aspirations, struggle, conflict, negotiation and cooperation among community members, other development agents, and a range of public and private agents that influence community life.

Finally, social innovation is inherently cultural. It is a quest for cultural recognition and identity in the economic, social and political lifeworlds; it is also often stirred by cultural dynamics – reacting to the alienation of citizenship, identity, social networks and so forth (negative motor), or unrolling potential for emancipation, capacity building, improved human rights and suchlike (positive motor). Within these dynamics, archetypes of social innovation such as bottom-up democracy for all, needs-revealing strategies, community built on diversity, and so on serve as references for a renewed cultural identity building process.

PART I
Social Innovation:
Needs Satisfaction, Community
Empowerment and Governance

Chapter 1

Social Innovation: Institutionally Embedded, Territorially (Re)Produced

Frank Moulaert

Introduction

'Social innovation' is a concept significant in scientific research, business administration, public debate and ethical controversy. As we will see in the next section, the term is not new, especially in the scientific world. But it has returned to prominence in the last 15 years, after a period of neglect. It is used in ideological and theoretical debates about the nature and role of innovation in contemporary society (Hillier et al. 2004), either to confront mainstream concepts of technological and organizational innovation, or as a conceptual extension of the innovative character of socio-economic development. That is, the concept enlarges the economic and technological reading of the role of innovation in development to encompass a more comprehensive societal transformation of human relations and practices (Moulaert and Nussbaumer 2008).

A variety of life-spheres and academic disciplines have taken on board the concept of social innovation. To begin with, social innovation is a hot topic in business administration where it refers to two new foci. The first one gives more attention to the social character of the firm: the firm as a network of social relations and as a community in which technological and administrative changes are just one part of the innovation picture, the institutional and social being of at least equal importance. To put it more strongly: the business administration literature increasingly stresses how many technological innovations fail if they are not integrated into a broader perspective in which changes in social relations within, but also embedding, the firm play a key role. If this sounds like the ultimate form of capitalism, that is, the commodification of all social relations within and across firms, it also refers to a second concern, which is to let firms play a more active social role in society – discursive or real. This sought-for social role often reflects a pure marketing strategy in the sense of 'make the firm look more socially responsible so as to sell better'; but it can also stand for a real alternative, ranging from a diversity of 'Corporate Social Responsibility' initiatives to the establishment of new units or subsidiaries that are fully active in the social economy, or/and have resolutely opted for ecologically and socially sustainable outputs and production models (Moulaert and Nussbaumer 2008). But social innovation is not only back on stage in business administration, it is the driving force of many NGOs, a structuring principle of social economy organizations, a bridge between emancipating collective arts initiatives and the transformation of social relations in human communities.

This edited volume is about social innovation and territorial development. It focuses on social innovation not only within a spatial context, but also as 'transformer' of spatial relations. It defines social innovation as the satisfaction of alienated human needs through the transformation of social relations: transformations which 'improve' the governance systems that guide and regulate the allocation of goods and services meant to satisfy those needs, and which establish new governance structures and organizations (discussion fora, political decision-making systems, firms, interfaces, allocation systems, and so on). Territorially speaking, this means that social innovation involves, among others, the transformation of social relations in space, the reproduction of place-bound and spatially exchanged identities and culture, and the establishment of place-based and scale-related governance structures. This also means that social innovation is quite often either locally or regionally specific, or/and spatially negotiated between agents and institutions that have a strong territorial affiliation.

Before focusing, in the third section of this chapter, on social innovation in and through space, I first adopt a more historical perspective and examine how the concept of social innovation has been present in academic literature since the beginning of the twentieth century, and even before.

Historic Antecedents of the Theory and Practice of Social Innovation

The concept of social innovation is not new. As far back as the eighteenth century, Benjamin Franklin evoked social innovation in proposing minor modifications within the social organization of communities (Mumford 2002), and in 1893, Emile Durkheim highlighted the importance of social regulation in the development of the division of labour which accompanies technical change. Technical change itself can only be understood within the framework of an innovation or renovation of the social order to which it is relevant. At the start of the twentieth century, Max Weber demonstrated the power of rationalization in his work on the capitalist system. He examined the relationship between social order and innovation, a theme which was revisited by philosophers in the 1960s. Amongst other things, he affirmed that changes in living conditions are not the only determinants of social change. Individuals who introduce a behaviour variant, often initially considered deviant, can exert a decisive influence; if the new behaviour spreads and develops, it can become established social usage. In the 1930s, Joseph Schumpeter considered social innovation as structural change in the organization of society, or within the network of organizational forms of enterprise or business. Schumpeter's theory of innovation went far beyond the usual economic logic, and appealed to an ensemble of sociologies (cultural, artistic, economic, political, and so on), which he sought to integrate into a comprehensive social theory that would allow the analysis of both development and innovation.

Finally, in the 1970s, the French intellectuals of the '*Temps des Cerises*' organized a debate of wide social and political significance on the transformation of society, and on the role of the revolts by students, intellectuals and workers. At the same time, a major part of the debate was echoed in the columns of the journal *Autrement*, with contributions from such prominent figures as Pierre Rosanvallon, Jacques Fournier and Jacques Attali. In their book on social innovation, *Que sais-je?*, Chambon, David and Devevey (1982) build on most of the issues highlighted in this debate. This 128-page book remains the most complete 'open' synthesis on the subject of social innovation to this day. In brief, the authors examine the relationship between social innovation and the pressures bound up within societal changes, and show how the mechanisms of crisis and recovery both provoke and accelerate social innovation. Another link established by Chambon et al. concerns social needs and the needs of the individual, individually or collectively revealed. In practice, social innovation signifies satisfaction of specific needs thanks to collective initiative, which is not synonymous with state intervention. According to Chambon et al., in effect the state can act, at one and the same time, as a barrier to social innovation and as an arena of social interaction provoking social innovation from within the spheres of state or market. Finally, these authors stress that social innovation can occur in different communities and at various spatial scales, but is conditional on processes of consciousness raising, mobilization and learning.

The authors cited up to this point cover the most significant dimensions of social innovation. Franklin refers to 'one-off' innovation in a specific context; Weber and Durkheim emphasize changes in social relations or in social organization within political and economic communities; Schumpeter focuses on the relationship between development and innovation where strong technical economic innovation is considered of prime importance and where the entrepreneur is viewed as a leader who, despite facing many difficulties, is able to introduce innovation into modes of societal organization. Most of these highlight the importance of social innovation within diverse types of institutions and institutional dynamics (such as public administration, world politics, enterprise, local communities, intergroup or community relations). Finally, Chambon et al. add to these dimensions by introducing the relationships between social needs and their individuation, societal change, and the role of the state. They thus offer a fuller picture of social innovation which provides a platform for global discussion on this theme.

Today's return to social innovation as a theme for research and as a principle structuring collective action is not at odds with the 'founding writings' described above. In tune with Schumpeter's work, in contemporary business literature, social innovation shows itself through the activities of the innovating entrepreneur who alters the social linkages at the core of the enterprise, to improve its functioning, to transform it into a social undertaking or to introduce a social rationale (for example see Manoury 2002, 5). Schumpeter and Weber are cited regularly by authors seeking to legitimize social transformation in organizational structures, in both business and public administration, where principles of social innovation are actively applied (for a survey see Moulaert and Nussbaumer 2008, Chapter 3).

Following on from the re-reading of the works of Benjamin Franklin, who perceived social innovation as the solution to specific life problems (Mumford 2002), and of the foundational writings of sociology, social innovation today can also be rediscovered within the artistic world, in which society and its structures can be creatively rethought. In effect, the arts re-invent themselves as sociology, as in the 'Sociologist as an Artist' approach, which underlines the importance of sociology as the science of innovation in society (Du Bois and Wright 2001). Finally, 'the return of social innovation', both in scientific literature and political practice, is demonstrated by the use of the concept as an alternative to the logic of the market, and to the generalized privatization movement that affects most systems of economic allocation; it is expressed in terms of solidarity and reciprocity (Liénard 2001; Nyssens 2000; Moulaert and Nussbaumer 2005b).

Social Innovation in Contemporary Social Science

In contemporary social science, there is growing interest in the idea of social innovation. I have singled out four spheres, or approaches, utilizing the concept which I present briefly here.

The first sphere is that of *management science* and its potential to share themes with other social science disciplines. For instance, within social science literature, authors emphasize opportunities for improving social capital which would allow economic organizations either to function better or to change; this would produce positive effects on social innovation in both the profit and non-profit sectors. This emphasis on and reinterpretation of social capital, which has also been taken on board in management science, would include economic aspects of human development, an ethical and stable entrepreneurial culture, and so forth, and thus facilitate the integration of broader economic agendas, such as those which advocate strong ethical norms (fair business practices, respect for workers' rights) or models of stable reproduction of societal norms (justice, solidarity, cooperation and so on) within the very core of the various entrepreneurial communities. However, the price paid for this sharing of the social capital concept across disciplines is that it has become highly ambiguous, and its analytical relevance is increasingly questioned (Moulaert and Nussbaumer 2005b).

The second sphere arises from the fields of *arts and creativity*. It encompasses the role of social innovation in social and intellectual creation. Michael Mumford unlocks this idea in a paper which defines social innovation as:

> l'émergence et la mise en œuvre d'idées nouvelles sur la manière dont les individus devraient organiser les activités interpersonnelles ou les interactions sociales afin de dégager un ou plusieurs objectifs communs. Au même titre que

d'autres formes d'innovation, la production résultant de l'innovation sociale devrait varier en fonction de son ampleur et de son impact. (2002, 253)

[the emergence and implementation of new ideas about how people should organize interpersonal activities, or social interactions, to meet one or more common goals. As with other forms of innovation, results produced by social innovation may vary with regard to breadth and impact.]

Mumford, author of several articles on social innovation in the sphere of arts and creativity, posits a range of innovations from the 'macro-innovations' of Martin Luther King, Henry Ford or Karl Marx to 'micro-innovations' such as new procedures to promote cooperative working practices, the introduction of new core social practices within a group or the development of new business practices (2002, 253). Mumford presents his own view of social innovation employing three main 'lines of work': the life history of notable people whose contributions were primarily in the social or political arena; the identification of capacities leaders must possess to solve organizational problems; the development, introduction and adaptation of innovations in industrial organizations. He then applies a mixed reading along these three lines to an examination of the work of Benjamin Franklin and arrives at a definition that parallels and shows synergies within the approach of the 'Sociologist as an Artist'.

The third sphere concerns social innovation in territorial development. Moulaert (2000) stresses local development problems in the context of European towns: the diffusion of skills and experience amongst the various sectors involved in the formation of urban and local development policies; the lack of integration between the spatial levels; and, above all, neglect of the needs of deprived groups within urban society. To overcome these difficulties, Laville et al. (1994) and Favreau and Lévesque (1999) put forward neighbourhood and community development models. Moulaert and his partners in the IAD project have suggested organizing neighbourhood development along the lines of the *Integrated Area Development* approach, (the *Développement Territorial Intégré*) which brings together the various spheres of social development and the roles of the principal actors by structuring them around the principle of social innovation. This principle links the satisfaction of human needs to innovation in the social relationships of governance. In particular, it underlines the role of socio-political capacity (or incapacity) and access to the necessary resources in achieving the satisfaction of human needs; this is understood to require participation in political decision making within structures that previously have often been alienating, if not oppressive (Moulaert et al. 2007). A similar approach has been proposed for regional development policy: the 'Social Region' model offers an alternative to the market logic of Territorial Innovation Models (TIM; see Moulaert and Sekia 2003), replacing it with a community logic of social innovation (Moulaert and Nussbaumer 2005a).

The fourth sphere in which social innovation is the order of the day is that of *political science and public administration*. Criticisms of the hierarchical character of political and bureaucratic decision making systems are well known and are at the root of new proposals concerned with change in the political system and, above all, in the system of public administration. Several approaches or initiatives have been developed: the use of territorial decentralization (regionalization, enlarging the power and competence base of localities) in order to promote citizen access to governance and government; an increase in the transparency of public administration; the democratization of administrative systems by promoting horizontal communication; a reduction in the number of bureaucratic layers. All are designed to give more control and influence to both users and other 'stakeholders' (Swyngedouw 2005; Novy and Leubolt 2005).

Social Innovation and Territorial Development

Social innovation analysis and practice have devoted particular attention to the local and regional territory. In Western Europe, but also in other 'post-industrial' world regions like North America and Latin America, urban neighbourhoods have been the privileged spatial focus of territorial development based on social innovation. There are many explanations for this focus. *First*, there is the high tangibility of decline and restructuring in urban neighbourhoods: plant closure in the neighbourhood or within its vicinity erodes the local job market; high density of low-income social groups manifests in spending behaviour and social interaction; lived experience of the consequences of physical and biotopical decline affects community life, and so on. Because of spatial concentration, in general, the social relations, governance dynamics and agents 'responsible for' the decline are more easily identifiable in urban neighbourhoods than in lower density areas or at higher spatial scales. Proximity feeds depression, fatalism, localized *déjà-vus*, and so on. But, *second*, spatial density simultaneously works as a catalyst for revealing alternatives, however meagre they may be; urban neighbourhoods spatially showcase the cracks of hope in the system (to paraphrase CityMine(d) which uses the term KRAX, or urban ruptures or crack lines – see KRAX Journadas n.d.). Their proximity to the institutional and economic arenas underscores the ambiguity of these neighbourhoods: they are both hearths of doom – they could not avoid or even 'architecture' the decline – and ambits of hope – these arenas of dense human interaction show and often become loci of new types of social relations and drivers of alternative agendas.

The ambiguity of the status of local territories as breeding grounds of socially innovative development is well known in the literature. On the one hand these territories very often have lived long histories of 'disintegration': being cut off from prosperous economic dynamics, fragmentation of local social capital, breakdown of traditional and often beneficial professional relations, loss of quality of policy delivery systems, and so on. In this context Moulaert and Leontidou (1995) have called such areas disintegrated areas (see also Moulaert 2000). On the other hand, several of these areas have been hosts for dynamic populations and creative migration flows which have been instrumental in (partly) revalorizing social, institutional, artistic and professional assets from the past, discovering new assets and networking these into flights towards the future. In this sense, there is an artificial split within the local community-based development literature between the more traditional 'needs satisfaction', 'problem solving' approach, and the more diversity-based, future-oriented community development approach which looks in particular at the identification of aspirations, strengths and assets of communities to move into a future of hope (see Chapter 2, by Gibson-Graham and Roelvink, in this book; Kretzmann and McKnight 1993).

A thesis defended throughout the chapters in this book is that needs satisfaction and assets for development approaches cannot be separated, either for the purpose of analysing local socio-economic development trajectories of the past, or for the construction of alternatives for the present and future. The philosophy of the Integrated Area Development approach is based on the satisfaction of basic needs in ways that reflect not only the alienation and deprivation of the past, but also the aspirations of the new future. This satisfaction should be effectuated by the combination of several processes:

- the revealing of needs, and of potentials to meet them, by social movements and institutional dynamics – within and outside the state sphere, with a focus, but a non exclusive focus, on the local scale;
- the integration of groups of deprived citizens into the labour market and the local social economy production systems (referring to activities such as housing construction, ecological production activities, social services);
- education and professional training leading to integration into the labour market, but also to more active participation in consultation and decision making on the future of the territory. The institutional dynamics should continually enrich local democracy, the relations with the local authorities and the other public as well as private partners situated outside the locality but taking part in the local development. The local community could in this way seek to regain control of its own governance, and put its own movements and assets at the heart of this process of renaissance (Martens and Vervaeke 1997; Mayer forthcoming; García 2006).

Looking more closely at how the above processes are materialized, Integrated Area Development is socially innovative in at least two senses. *First of all*, from a sociological perspective, IAD involves innovation in the relations between individuals as well as within and among groups. The organization of groups and communities, the building of communication channels between privileged and disfavoured citizens within urban society, the creation of a people's democracy at the local level (neighbourhood, small communities, groups of homeless or long term unemployed, and so on) are factors of innovation in social relations. Governance relations are a part of the social relations of Integrated Area Development; without transformation of institutions and practices of governance, it becomes more or less impossible to overcome the fractures caused by different disintegration factors within communities and their local territories (Garcia 2006; LeGalès 2002).

The *second meaning* of social innovation within IAD reinforces the first: it evokes the 'social' of the social economy and social work (Amin et al. 1999). The challenge here is to meet the fundamental needs of groups of citizens deprived (*démunis*) of a minimum income, of access to quality education and other benefits of an economy from which their community has been excluded. There are different opinions on the nature of fundamental or basic needs, but a consensus is developing that a *contextual* definition is needed, according to which the reference 'basket' of basic needs depends on the state of development of the national/regional economy to which a locality belongs. 'State of development' here refers to the income per capita, the distribution of income and wealth and the cultural dynamics and norms determining so-called secondary needs.

The combination of these two readings of social innovation stresses the importance of creating 'bottom-up' institutions for participation and decision-making, as well as for production and allocation of goods and services (see Figure 1.1). The mobilization of political forces which will be capable of promoting integrated development is based on the empowerment of citizens deprived of essential material goods and services, and of social and political rights. Such a mobilization should involve a needs-revealing process different from that of the market, which reveals only necessities expressed through a demand *backed up by purchasing power* – the only demand that is recognized in orthodox economics. In a decently working Welfare State/economy persons and groups without sufficient purchasing power could address themselves to the existing systems of social assistance and welfare for the satisfaction of their needs. But these sources of goods and services are often downsized by the austerity policy of the neoliberal state or by the dominance of allocation criteria based on individual merits; they therefore do not always provide an acceptable level or quality.

Experiences of alternative territorial development, inspired and/or steered by socially innovative agencies and processes, unveil different aspects of the double definition of social innovation at the level of cities and urban neighbourhoods. Professional training targets the reintegration of unemployed into the regular labour market but also into new production initiatives in the construction sector, the consumption goods sector, ecological production activities, and so on (Community Development Foundation 1992). In many localities, new networks for production, training and neighbourhood governance are being explicitly constructed (Jacquier 1991; OECD-OCDE 1998; Favreau and Levesque 1999; Fontan et al. 2004; Drewe et al. 2008). But to achieve the ambitions of Integrated Area Development, the different pillars of IAD (territorially based needs satisfaction, innovation in social relations and socio-political empowerment) should be effectively materialized and connected. Far from seeking to impose an 'integral integration', connecting *all* the theoretical constituents of the approach, we consider territorial development projects as integrated when at least two 'sectors' (sectors of materialized IAD pillars are: training and education, labour market, employment and local production) are linked and when an active governance (reproduced through community empowerment and institutional dynamics) steers or feeds this connection (Moulaert 2000). Socially innovative governance in IAD has as an objective the democratization of local development, through activating local politics and policy-making, simplifying the functioning of institutions and attributing a more significant role to local populations and social movements (Novy and Leubolt 2005). The empowerment of the local population is primordial to democratic governance and the building of connections between the sections of the local system. It is, in the first place, implemented by jointly designed procedures of consultation and shared decision making about the needs to be revealed and met, and about the assets that could be put on track to this end.

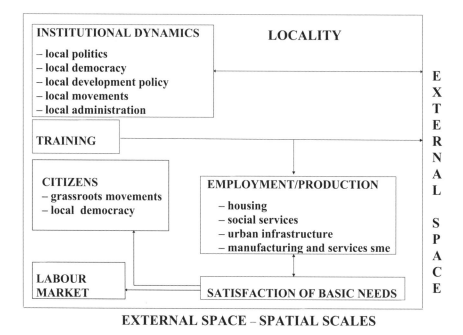

EXTERNAL SPACE – SPATIAL SCALES

Figure 1.1   Social innovation and integrated area development
Source: Based on Moulaert et al. (2000).

The Social Relations of Territorial Community Development

There exist many different orientations for strategies of social innovation at the level of neighbourhoods and localities (cultural, technological, artistic, artisanal; and equitable provision of 'proximity services' – see *City* 2004; André et al. Chapter 9 in this book). This book (especially the second part) focuses on territorially integrated experiences or projects that combine various initiatives built on forces that are socially organized at diverse but articulated spatial scales, with the purpose of satisfying the existential needs of inhabitants, and in the first place those inhabitants deprived of resources.[1] The rich diversity of research into such initiatives allows exploration of the relationship between path dependence, the present and the future of neighbourhoods, as well as between the analysis of and the strategies for territorial and community development. These relationships are difficult and refer as much to the problems raised by the (structural, institutional) determinants stemming from socioeconomic history as from the potential conflicts and opportunities that the confrontation of 'past' and 'future' as well as 'here' and 'elsewhere' can generate. In this respect, the analysis of path dependency as embedded in territorial development helps to avoid a deterministic reading of both the past and the structural–institutional context in which territorial and community development (should) take(s) place.

Thus considered, the '*va et vient*' between lived development and pro-active development, has generated a number of observations on the nexus of social relations and territorial development:

- The social relations of territorial development are not legible in general terms, but require an explication of the nature of development, the type of socio-political development, the nature of the strategic actors and the relationships with the territory – in all its social, political, economic, etc. dimensions.
- The same holds for the analysis of social capital within territorial social relations, where one should avoid at any price an instrumental interpretation. Social capital is socially embedded – and this is not a tautological observation but rather a confirmation of the fragmented nature of social relations and their links with the economic, cultural and symbolic capital of individuals and groups that belong to specific social communities (Moulaert and Nussbaumer 2005b). From this viewpoint, social innovation means not only the (re)production of social capital(s) in view of the implementation of development agendas, but also their protection from fragmentation/segmentation, and the valorization of their territorial and communal specificity through the organization and mobilization of excluded or disfavoured groups and territories.

I conclude that social innovation in territorial development must be addressed through a detailed analysis of how social and territorial logics interact with each other. In Lefebvrian terms (1991 [1974]) one should indeed devote reflection to the following questions:

---

1   See also the work of Christian Jacquier, Jean-Louis Laville, Juan-Luis Klein, J.K. Gibson-Graham, Frank Moulaert, Pavlos Delladetsima, Serena Vicari Haddock, Jean-Cédric Delvainquière, Christophe Demazière and the EU project SINGOCOM (n.d.).

- How does social innovation relate to the social production of space?
- Should it only be interpreted in terms of production (and production of perceived space) or is it also part of conceived and lived space?

Within much of the literature, social innovation in its territorial dynamics is expressed in terms of the representation of space, or even of spatial practice. But in reality its materialization depends significantly on its relations with the lived space and its perception; in fact it is this lived space that will produce the images and the symbols to develop a new language, and the imagineering tools to conceptualize a future social space.

References

Amin, A., Cameron, A. and Hudson, R. (1999), 'Welfare as Work? The potential of the UK social economy', *Environment and Planning A* 31:11, 2033–51.

Chambon, J.-L., David, A. and Devevey, J.-M. (1982), *Les innovations sociales* (Paris: Presses Universitaires de France).

City (2004), special topic on Bruges Cultural Capital of Europe, 8:2.

Community Development Foundation (1992), *Out of the Shadows: local community action and the European Community* (Dublin: European Foundation for the Improvement of Living and Working Conditions).

Drewe, P., Klein, J.-L. and Hulsbergen, E. (eds) (2008), *The Challenge of Social Innovation in Urban Revitalization* (Amsterdam: Techne Press).

Du Bois, W. and Wright, R. (2001), *Applying Sociology: making a better world* (Boston, MA: Allyn and Bacon).

Favreau, L. and Lévesque, B. (1999), *Développement Economique Communautaire. Economie Sociale et Intervention* (Sainte-Foye: Presses Universitaires du Québec).

Fontan, J.-M., Klein, J.-L. and Tremblay, D.-G. (2004), 'Collective Action in Local Development: the case of Angus Technopole in Montreal', *Canadian Journal of Urban Research* 13:2, 317–36.

García, M. (2006), 'Citizenship Practices and Urban Governance in European Cities', *Urban Studies* 43:4, 745–65.

Gibson-Graham, J.K. and Roelvink, G. (2008), 'Social Innovation for Community Economies', in MacCallum et al. (eds).

Hillier, J., Moulaert, F. and Nussbaumer, J. (2004), 'Trois essais sur le rôle de l'innovation sociale dans le développement spatial', *Géographie, Economie, Société* 6:2, 129–52.

Jacquier, C. (1991), *Voyage dans dix quartiers européens en crise* (Paris: L'Harmattan).

KRAX Journadas 2.0 (n.d.), Autonomia (conference website) <http://krax-jornadas.citymined.org/index-eng.html>.

Kretzmann, P. and McKnight, J.L. (1993), *Building Communities from the Inside Out: a path toward finding and mobilizing a community's assets* (Evanston, IL: Institute for Policy Research).

Laville, J.-L., Gardin, L., Lévesque, B. and Nyssens, M. (1994), *L'économie solidaire, une perspective internationale* (Paris: Desclée de Brouwer).

Lefebvre, H. (1991 [1974]), *The Production of Space* (Oxford : Blackwell).

LeGalès, P. (2002), *European Cities: social conflicts and governance* (Oxford: Oxford University Press).

Liénard, G. (ed.) (2001), *L'insertion: défi pour l'analyse, enjeu pour l'action* (Mardaga: Sprimont).

MacCallum, D., Moulaert, F., Hillier, J. and Vicari Haddock, S. (eds) (2008), *Social Innovation and Territorial Development* (Aldershot: Ashgate).

Manoury, L. (2002), *L'entrepreneur social et l'enjeu de sa professionalisation* (Aix-en-Provence: Université Coopérative Européenne).

Martens, A. and Vervaeke, M. (eds) (1997), *La polarisation sociale des villes européennes* (Paris: Anthropos).

Mayer, M. (forthcoming; 2010), *Urban Social Movements* (Oxford: Blackwell).

Moulaert, F. et al. (2000), *Globalization and Integrated Area Development in European Cities* (Oxford: Oxford University Press).

Moulaert, F. and Leontidou, L. (1995), 'Localités désintégrées et stratégies de lutte contre la pauvreté', *Espaces et Sociétés* 78, 35–53.

Moulaert, F., Martinelli, F., Gonzalez, S. and Swyngedouw, E. (2007), 'Introduction: Social Innovation and Governance in European Cities. Urban development between path-dependency and radical innovation', *European Urban and Regional Studies* 14:3, 195–209.

Moulaert, F. and Nussbaumer, J. (2005a), 'The Social Region: beyond the territorial dynamics of the learning economy', *European Urban and Regional Studies* 12:1, 45–64.

Moulaert, F. and Nussbaumer, J. (2005b), 'Defining the Social Economy and its Governance at the Neighbourhood Level: a methodological reflection', *Urban Studies* 42:11, 2071–88.

Moulaert, F. and Nussbaumer, J. (2008), *La logique spatiale du développement territorial* (Sainte-Foye: Presses Universitaires du Québec).

Moulaert, F. and Sekia, F. (2003), 'Territorial Innovation Models: a critical survey', *Regional Studies* 3, 289–302.

Mumford, M.D. (2002), 'Social Innovation: ten cases from Benjamin Franklin', *Creativity Research Journal* 14:2, 253–66.

Novy, A. and Leubolt, B. (2005), 'Participatory Budgeting in Porto Alegre: the dialectics of state and non-state forms of social innovation', *Urban Studies* 42:11, 2023–36.

Nyssens, M. (2000), 'Les approches économiques du tiers secteur', *Sociologie du travail* 42, 551–65.

OECD-OCDE (1998), *Intégrer les quartiers en difficulté* (Paris: OCDE, Développement Territorial).

SINGOCOM (n.d.), project website < http://users.skynet.be/bk368453/singocom/index2.html>.

Swyngedouw, E. (2005), 'Governance Innovation and the Citizen: the Janus face of governance-beyond-the-state', *Urban Studies* 42:11, 1991–2006.

Chapter 2

Social Innovation
for Community Economies

J.K. Gibson-Graham and Gerda Roelvink

Beginning a Conversation for Change

The Community Economies project is an ongoing effort to contribute to an emerging economic politics, one that is centred on the practice of economic self-determination, oriented by the vision that 'another world is possible' and committed to postcapitalist economic futures (see Community Economies n.d.; Gibson-Graham 2006). The project seeks to 'reclaim the economy' as a situated and diverse space of ethical decision making and negotiated interdependence; through that process of reclamation, the economy – a remote and powerful sphere that seems to dictate our lives – becomes instead a familiar, even intimate, space of engagement.

Central to the project are three key elements: (1) rethinking economy to create a conceptual platform for its reenactment; (2) enrolling and resubjecting communities and individuals (including ourselves) in new worlds of possibility; and (3) promoting collective action to build community economies. A community economy is not defined by geographic or social commonality; it is an ethical and political space of decision making in which interdependence is constructed as people transform their livelihoods and lives.

The Community Economies project is one of many contemporary projects seeking to foster new worlds and innovative economies. In this paper, we attempt to open a conversation between Community Economies and another such project, a multiphase European-based research initiative on local social innovation. This initiative is variously designated ALMOLIN and SINGOCOM (alternative models for local innovation and social innovation in local community governance), but for the sake of clarity we will call it simply the Local/Social Innovation project. Like Community Economies, the Local/Social Innovation project has affinities with the politics of possibility of the World Social Forum, and highlights the ways in which social movements are starting from where they are in order to meet previously unmet needs (see Moulaert et al. 2002). In particular, Local/Social Innovation is seen as a way to include marginalized groups within social and political governance institutions and processes. Local/social innovators privilege participation of the needy themselves in projects that seek to harness resources in the face of contemporary economic and social crises.

So why might a conversation between these two projects be of interest or use? In *A Postcapitalist Politics*, J.K. Gibson-Graham ponders the proliferation of a diverse range of economic initiatives around the globe and asks '… how do we multiply, amplify, and connect these different activities? How do we trace "connections between diverse practices … to dissolve the distinctions between inside and outside the movement"… and thus actualize movement goals in a transformed social order?' (2006, 80–81). In beginning a conversation between two different, but related, bodies of research, this chapter offers one way of responding to the challenge of connection. While the Local/Social Innovation project is oriented towards studying and theorizing movements and organizations, Community Economies is more interested in instituting and performing them, primarily through action research. Nevertheless, both projects can be seen as promoting and supporting an economic and social order that is emergent but not fully constituted. By making this nascent and largely non-credible 'movement' an object of research, and by enhancing the knowledge and self-knowledge of the projects and subjects involved, they participate in bringing it into being as a transformative force worldwide.

As an academic exercise, this somewhat one-sided conversation necessarily comes out of a critical tradition, one from which we would hope to step aside, at least for the moment. In the spirit of the new politics that seeks alliances rather than mergers (or deeper divisions), this chapter inquires about the ways in which Local/Social Innovation connects and overlaps with the project of creating innovative and resilient community economies. In doing so, it develops four key themes:

- acceptance of diversity as the starting point
- local ethics of individual and social needs
- community governance
- the building of innovative communities.

Each of these themes resonates with a core principle of the politics of 'other worlds'. In the concluding section of the chapter, we highlight fundamental similarities and differences between the two projects that make them potentially fruitful contributors to each other and to the self-reflection of each.

Starting with Diversity: A Project of Visibility

It may come as no surprise that both Community Economies and Local/Social Innovation are grounded in an appreciation of social and economic diversity. Yet for both frameworks, the recognition of diversity is not simply a matter of reflecting the truth of the world; rather it is a strategic theoretical choice. For Gibson-Graham (2006), for example, identifying and describing a rich diversity of economic practices and organizations is a key aspect of a politics of language – making space for, and giving legitimacy to, forms of social and economic activity that are obscured and devalued by a capitalocentric worldview. By bringing these to the attention of participants and stakeholders, they hope to widen the field of possibility for economic activism and development.

In a similar vein, Moulaert and his colleagues are working to broaden and diversify the dominant notion of innovation, which is limited to innovations that enhance economic efficiency (Moulaert et al. 2005, 1973; Moulaert and Nussbaumer 2005). They define social innovation more broadly in terms of the inclusion of the marginalized into a range of areas, including education systems, labour markets, political institutions and sociocultural life (Moulaert et al. 2005, 1970). In their research, they have focused primarily on innovative ways of meeting the needs of excluded groups and on innovative governance structures in organizations created by and/or serving those groups. Taken together, such organizations are understood as constituting the social economy, a concept that encompasses a wide array of initiatives oriented toward the satisfaction of needs (sometimes called the third sector, or the solidarity economy) (Moulaert and Ailenei 2005). The development of the social economy is seen as a response to the distributive failures of market and state; its presence has thus fluctuated over time with the business cycles of national and global economies, changing labour markets, the impact of World Wars, the emergence and decline of the Welfare State, and other events (Moulaert et al. 2005). Depending on the historical and geographical context, the social economy will take diverse forms and may include initiatives that draw on the market and the state as a way to satisfy needs that are unmet by these institutions (Moulaert and Ailenei 2005).

SINGOCOM (centred on 'social innovation in governance for local communities') documented diverse social economy projects in Europe with the aim of providing 'these initiatives with a new synthesis of theoretical foundations' (Moulaert et al. 2005, 1970). Initiatives in the SINGOCOM database include a mediating and coordinating neighbourhood organization in Berlin; an organization to support skill sharing and the development of cooperatives in Sunderland, UK; an informal social support network in Quartieri Spagnoli, a poor section of Naples; and arts projects to record local histories in Cardiff (Moulaert et al. 2005, 1970). All of these innovative projects are, at the same time, projects of inclusion, whether geared towards meeting the material needs of the marginalized, opening social arenas to the previously excluded, or giving 'voice' to those who have had little or no say in political life.

Community Economies is likewise interested in strategies for including the marginalized by developing diversified social economies. Their strategic entry point, however, is a reconceptualization of the *entire* economy as a diverse social arena, creating an alternative economic language of the 'diverse economy'. This language provides a discursive space and an open-ended set of categories with which to make visible the wide range of transactions, forms of labour, and economic organization that have been marginalized by the discourse of the 'capitalist economy' (see Figure 2.1). The representation of economy in Figure 2.1 both ruptures the presumed unity of capitalism and calls into question its presumptive dominance, especially when we recognize that noncapitalist market and nonmarket activity constitutes well over 50 per cent of all economic activity (Gibson-Graham 2006, 68). In particular, unpaid labour in households and neighbourhoods constitutes 30 to 50 per cent of economic activity in both rich and poor countries (Ironmonger 1996). Interestingly, the social economy of Local/Social Innovation enrols many of the practices and organizational forms of the diverse economy, including alternative market and non-market transactions, alternatively paid and unpaid labour, and alternative capitalist and non-capitalist enterprises.

Like SINGOCOM, the diverse economies framework is a tool for inventorying and describing economic diversity in an open way that begins rather than forecloses discussion. In Community Economies action research around the world, mapping the local diverse economy has allowed people to see and valorize the economic activity in which they are already engaged. Whereas conventional economic development usually starts with the presumption that a community is lacking in and therefore needs capitalist development, the Community Economies project presumes the opposite; it affirms the presence of hidden assets and capacities that

Transactions	*Labour*	*Enterprise*
MARKET	**WAGE**	**CAPITALIST**
ALTERNATIVE MARKET *Sale of public goods* *Ethical 'fair-trade' markets* *Local trading systems* *Alternative currencies* *Underground market* *Co-op exchange* *Barter* *Informal market*	*ALTERNATIVE PAID* *Self-employed* *Cooperative* *Indentured* *Reciprocal labour* *In kind* *Work for welfare*	*ALTERNATIVE CAPITALIST* *State enterprise* *Green capitalist* *Socially responsible firm* *Non-profit*
NON-MARKET *Household flows* *Gift giving* *Indigenous exchange* *State allocations* *State appropriations* *Gleaning* *Hunting, fishing, gathering* *Theft, poaching*	*UNPAID* *Housework* *Family care* *Neighbourhood work* *Volunteer* *Self-provisioning labour* *Slave labour*	*NON-CAPITALIST* *Communal* *Independent* *Feudal* *Slave*

Figure 2.1   A diverse economy

could provide a useful starting place for previously unimagined development paths. In the Latrobe Valley in Australia, the municipality of Jagna in the Philippines, and western Massachusetts in the United States, Gibson-Graham (2006) and their colleagues have shown how mapping assets and capacities rather than needs and deficiencies reveals that there is something to build upon and opens up discussions of the direction that a building process might take. In these community discussions, there is surprise and relief at discovering existing and potential alternatives to the vision of capitalist development on offer from governments, international institutions, and many academics and NGOs.

Shared principle for 'other worlds': making diversity visible, promoting credibility, enhancing possibility.

The Local Ethics of Individual and Social Needs: A Vision of Distributive Justice

Both Community Economies and Local/Social Innovation suggest that social inclusion through innovative community initiatives can be understood as an ethical practice of locality (Gibson-Graham 2003). In the Local/Social Innovation framework, inclusion is referenced to the needs of the excluded, seen as social groups that are in some way marginalized from society, from immigrants to the disabled. The project brings the unmet, and often unregistered, needs of marginalized groups into full view as a trigger for local/social innovation.

Although needs are recognized to be diverse and changing, they can be summed up as 'alienated basic needs', and the project itself concentrates on needs for access to resources and political participation (Moulaert et al. 2005, 1976). Needs are viewed as both individual and social. Thus Local/Social Innovation sees personal desires for dignified livelihoods and political voice as integrated with community development strategies. Projects of Integrated Area Development, for example, initiate and support community enterprises that improve individual living conditions and, at the same time, strengthen the local economy and its social, cultural and physical infrastructure (Moulaert et al. 2000).

Social Enterprise Sunderland, one of the case studies in the SINGOCOM project, exemplifies the ways in which cooperatives can be established to meet social and individual needs. Social Enterprise Sunderland developed from a joint venture responding to local concerns about individual needs for employment and dignity in work, as well as social needs for community wealth in the form of infrastructure and housing. They offer assistance to groups wishing to develop and build cooperative and social enterprises and are beginning to link these enterprises in broader networks. They also promote community development by managing community projects, such as a co-op centre, a sports centre and a community primary school (Moulaert et al. 2005, 1970; Social Enterprise Sunderland n.d.).

Viewed through the lens of the Community Economies project, the range of needs identified by Local/Social Innovation implies the possibility of ethical debate and local decision making about what is necessary for individual and social life, and how the common wealth (the 'commons') that defines a community is to be shared. Such ethical decision making is both the marker of community economies and the process by which they are formed; at the most fundamental level, it is both a practice and an enactment of 'being-in-common', which is Jean-Luc Nancy's phrase for the radical and unavoidable commonality of coexistence with others (Nancy 1992). For Gibson-Graham and her colleagues, a community economy is an ethical practice of being-in-common, a space of negotiated interdependence where decisions are made about what is necessary for individual and social life, and the 'question of how to live together' is openly engaged (2006, 81–2).

Gibson-Graham uses the Mondragón Cooperative Corporation in the Basque region of Spain to exemplify the kind of ethical debate that is constitutive of community economies. Reflecting on the 50-year history of the Mondragón cooperatives, which were established by the local Basque people to provide employment and allow for economic self-determination, she emphasizes how the surplus generated by each cooperative was pooled through the cooperative bank in order to capitalize more cooperatives and expand the community economy of Mondragón. The decision to use the surplus in this way and to keep wages equivalent to those of other workers in the Basque region established a certain standard of living for Mondragón workers; whatever was declared to be surplus was not available to individual workers to increase their personal and family consumption. In Mondragón, what is necessary and what is surplus are neither given by nature nor decreed by a capitalist employer; they are constituted relationally by the cooperators themselves, in the ethical process of balancing their individual desires for consumption with their goals for the Basque people and the local economy (Gibson-Graham 2006).

From a Community Economies perspective, cooperative enterprises do not simply respond to unmet needs but provide a site in which such needs are denaturalized and opened up for discussion. In the very different context of the labour movement, we can see a similar process of denaturalization and the negotiated status of necessity; as the movement has struggled to increase workers' wages and improve working conditions, workers have redefined what is necessary for a fair and decent way of life (Gibson-Graham 2006, 89). Needs are also negotiated and redefined in the process of establishing differential taxation, through which wealth is collected from some individuals and redistributed to others on the basis of a malleable and changing vision of what is necessary to support human existence. Indeed, ethical decisions around needs are made in all sorts of contexts, signalling the presence of community economies (or aspects of them) in unexpected places and at a variety of scales.

Shared principle for 'other worlds': meeting needs directly, innovatively, democratically – distributive justice on the agenda.

Community Governance: Cultivating New Practices and Subjects

From the perspective of Local/Social Innovation, needs satisfaction requires changes in social relations and, in particular, relations of governance. The development of governance capacity, or the 'ability of the institutional relations in a social milieu to operate as a collective actor' (Moulaert et al. 2005, 1984), is thus a focus of concern for the project, which has a particular interest in the ways in which socially innovative governance initiatives emerge from the grassroots and take hold in a wider context (Gonzalez and Healey 2005). When used as an 'analytical tool' in Local/Social Innovation, a focus on governance draws attention to the people involved in decision making and also to the forms and flavours of such decision making. To resist prevailing modes of thinking and acting, innovative governance must involve a range of people, including 'non-traditional actors' (Gonzalez and Healey 2005, 2061). The concern for social innovation in governance also suggests that the success of innovative enterprises is not simply to be measured by their life span and growth but also by the 'seeds' and 'sediments' that may influence future practice (2005, 2065). Innovation in governance is actually a useful metric for evaluating social enterprises, as the scope of success and failure extends beyond quantifiable outcomes of particular projects to more general changes in participation, practices and values.

The SINGOCOM project examines the Ouseburn Trust in Newcastle upon Tyne as an example of innovative governance. The Trust was initiated by local church and community leaders in response to development plans for the Ouseburn Valley that threatened the local commons, particularly the natural environment and industrial heritage. Over time the Trust developed networks with other local initiatives and a relationship with government, relationships that were formalized and institutionalized through the creation of an Advisory Committee. Gonzalez and Healey draw on the story of the Ouseburn Trust to highlight several aspects of innovative governance. These include the role of 'non-traditional actors' such as church leaders; partnerships with other enterprises and government, which involve adaptation by innovators to formal governance rules and processes as well as the exchange of ideas; and a more general change in ideas and values of governance institutions through the inclusion of diverse participants and areas of concern (Gonzalez and Healey 2005).

In the Community Economies approach, governance has been framed primarily in terms of subjectivity and subject formation, particularly the ethical practice of self-transformation that is involved in producing subjects for a community economy. The understanding that the economy is something we do, rather than something that does things to us, does not come naturally or easily. Innovative economic subjects must be nurtured and cultivated to value and act upon their interdependence.

The experience of Argentina during the early years of this century offers an inspiring example of self-cultivation as an aspect of innovative governance emerging from the grassroots. When hundreds of thousands of Argentineans became unemployed because of the economic crisis, the unemployed started to build community economies by engaging in barter and using alternative currencies, providing neighbourhood-based social services and schooling, and taking over factories and running them cooperatively. But they had to remake themselves in order to do this. To transform themselves into community economic subjects, they created a cooperative radio station; they went to the World Social Forum in Porto Alegre to see themselves reflected in others who were also engaged in projects of self-determination; they opened a school to teach themselves how to make their own history. Gibson-Graham has called this deliberate process of self (and other) transformation 'a politics of the subject' (2006), but it could just as easily be seen as a mode of (self) governance, oriented toward the creation of 'other worlds'.

Community Economies is also interested in governance processes within social and community enterprises. In order to deliver on the dual objectives of benefiting those involved in the enterprise and also the wider community, community enterprises require novel governance strategies as well as innovative ways to evaluate economic and social performance. At present, the Community Economies Collective is collaborating with community enterprises in the United States, the Philippines and Australia to produce a template for self-study for community enterprises. In this project, the governance strategies of community enterprises are being treated as experimental moves, that is, as innovations to be learned from both for the enterprises involved and for nascent or future community enterprises. One outcome of the project will be an alternative metric of success (or failure) that will clearly distinguish community enterprises from the mainstream enterprises against which they are often measured and found insufficient. This effort at self-knowledge and self-evaluation can be seen as helping to bring the community enterprise sector into the next stage of being – recognized by itself and others (including planners and policymakers) as a crucible of social innovation and an important sector of the economy, with its own distinctive dynamics, modes of governance and criteria of success.

> Shared principle for 'other worlds': self-determination and innovative governance at every scale, from self to world.

Building Innovative Communities: A New Economic Politics

Both Local/Social Innovation and Community Economies are interested in projects concerned with building new communities in which innovative governance is a central feature. In particular, scholars of Local/Social Innovation view the inclusion of marginalized groups into the 'politico-administrative system' as a key 'political rationale' in the diverse initiatives they study (Moulaert et al. 2005, 1970). The participatory budgeting practised in Porto Alegre, Brazil, exemplifies the type of political empowerment they seek to understand and promote.

Andreas Novy and Bernhard Leubolt analyse the Porto Alegre experience by placing it in the context of the tension between democracy and capitalism in Latin America. From their perspective, participatory budgeting aims to strengthen the democratic state and, at the same time, contribute to an alternative economy by including local people in decision making about the distribution of public money. In order to work well, it requires a sizeable and decentralized budget and substantial public participation (Novy and Leubolt 2005, 2026–7).

Novy and Leubolt's research demonstrates that participatory budgeting has widespread benefits. In Porto Alegre, it has greatly increased transparency in budgetary decisions and has installed democratic processes, including both direct democracy and the election of representatives to participate in ongoing decision making. It has also provided a process for achieving distributive justice. Like other projects collected under the banner of local/social innovation, participatory budgeting responds to local needs and tends to benefit the needy more than the well-off (Novy and Leubolt 2005, 2028). In Porto Alegre, it has increased the civic participation of the socially marginalized and provided public assistance to a diverse array of small projects.

Participatory budgeting has had another interesting outcome in Porto Alegre. As Novy and Leubolt (2005, 2030–31) describe it, the process prompts participants to articulate their individual needs in relation to community needs. Open political contestation facilitates the re-expression of 'my' needs as 'our' needs; in moving from 'I' to 'we', participants develop and express a sense of themselves as members of a community. They become, in the language of Community Economies, communal economic subjects – open to new forms of association and to the individual becomings that arise when connection is forged and interdependence acknowledged.

In action research in the Latrobe Valley of southeastern Australia and the Pioneer Valley of western Massachusetts in the United States, the Community Economies project found that building community economies entails an ongoing process of cultivating subjects who can open up to new forms of economic being (Gibson-Graham 2006). Initially this required working closely with action research participants to elicit their painful attachments to the dominant economy, giving them space to air their sense of economic injury and deficiency. As participants became involved in inventorying and representing a diverse and surprisingly vibrant local economy, their existing (narrow and relatively powerless) economic identities were destabilized, and they began (tentatively at first) to experience their economic selves in very different ways. Rather than feeling needy and deficient, isolated in an environment of scarcity, they could see themselves as having assets and capacities, embedded in a space of relative abundance. As the research process continued, revealing innovative livelihood strategies, informal transactions and a wealth of caring connections, a sense of possibility – again, tentatively expressed – became palpable among the participants: some were motivated to take up new activities; others were able to revalue old ones, not formerly seen as economic; and still others became involved in community enterprises, showing their willingness to relate to people in unfamiliar ways (Gibson-Graham 2006, Ch. 6).

In addition to the project of opening themselves and others to economic possibility, the Community Economies project is engaged in studying and fostering collective action to construct community economies on the ground. Rather than laying out the contours of an ideal community economy (which might obscure or preempt the decisions of communities themselves), they offer four 'coordinates' – necessity, surplus, consumption and commons – around which interdependence could be negotiated and explored (Gibson-Graham 2006, Chs 4 and 7). These coordinates constitute a rudimentary lexicon of interdependence, and collective decision making around them can be understood as the ethical 'dynamics' of a community economy.

Constructing the vision of a community economy around the four coordinates highlights the interdependencies among what are usually targets of single-issue politics (with the exception of surplus, which is not generally considered outside of cooperatives). The living wage movement, for example, concentrates on access to *necessities* of life; the simplicity movement devotes itself to lifestyle and *consumption*; the environmental movement focuses on protecting and restoring the *commons* in various forms. Bringing these issues together creates a complex field of decision in which trade-offs and other relationships can be examined and discussed – the question, for example, among a group of migrants, of whether to pool their (surplus) remittances and, if they decide to do so, how to use the money that results. Do they want to help replenish the fisheries at home, or to shore up the incomes of the elderly and disabled, or to invest in enterprise development so that migration will not be necessary for future generations? These questions open up different ways of consuming a common store of wealth, and beg the question of who should be involved in making the decisions. For the community involved, however constituted, the decisions taken will directly or indirectly affect a variety of forms of necessity, consumption and commons, not to mention the size of a possible future surplus. What the coordinates provide is a starting place for economic decision making that provisionally maps a complex ethical space, and creates potential connections between what are seemingly disparate and distant constituencies and issues.

> Shared principles for 'other worlds': new framings for economic politics; constituting 'we' in novel ways.

Conclusion

Situating Local/Social Innovation and Community Economies together in this chapter, we have highlighted areas of common concern and distilled them into widely shared principles for a politics of 'other worlds'. Yet it would be equally possible to see these two projects as very different, each with a well honed self-conception that separates and distances it from the other (and indeed from all other initiatives). For us, part of the ethical challenge of engagement with other projects lies in adopting an experimental rather than a critical stance; this means that differences are examined for what they can teach us, rather than presumed to be signs of deficiency on one or the other side. It is in this spirit of openness that we conclude the chapter with a discussion of salient differences between the two projects, grounded in a summary of their overlapping concerns.

Both projects are clearly concerned with redressing marginalization. Local/Social Innovation is primarily interested in marginalized social groups and their inclusion in social decision making (governance) and social allocation (meeting unmet needs). Community Economies conceives marginalization differently. They see everyone as marginalized by a dominant conception of economy that is assumed to govern itself and to disproportionately affect the surrounding social space. In the face of this general marginalization, their language politics attempts to bring widespread but discursively marginalized activities to light, revaluing caring labour, informal transactions, alternative enterprises (and more) as major economic forces, and thereby revaluing the subjects who enact them.

Turning around the emphasis on marginalization, both projects are concerned about fostering its opposite – social justice and inclusion. Local/Social Innovation is interested in the inclusion of the marginalized in democratically governed projects that address their needs, and in strategies of connecting these innovative projects with existing governance institutions. Community Economies focuses on rethinking economy as a social space of interdependence. Inclusion in this space requires cultivating new forms of self-recognition, through which individuals and groups come to see themselves as shaping/governing economic processes rather than as simply being subjected to them.

Perhaps the most interesting overlap between the two projects is their concern with theorizing the dynamics of innovation. Local/Social Innovation is interested in the growth and development of the social economy, which they see as emerging both from people's changing aspirations and also in response to the crises of a larger (capitalist) economy that periodically fails to meet people's needs for employment and well-being. In this latter framing, capitalism is positioned as the principal motor of change, with the social economy and social innovation seen as responding (Moulaert et al. 2005). At the same time, social innovation is (implicitly) understood as a present and proliferative force, ready to be called into action at critical moments; it emanates from the marginalized themselves as they struggle to have their needs met, including their needs for changes in social relations and governance.

While similarly concerned with dynamics, Community Economies is interested in displacing capitalism from the driver's seat of social and economic change. They turn their attention away from the so called structural dynamics of a capitalist system, and emphasize instead the ethical dynamics of decision making involved in constructing community economies. While they recognize and admire the capacity of the marginalized to spontaneously engage in social innovation, at the same time they are attuned to the resistances of those marginalized by the economy (all of us). Subjects tend to experience the economy as an external, almost colonizing power, to which they are beholden. Even within the domain of activism and entrepreneurship, the economic imagination is often timid when it comes to thinking outside the capitalist box. This means that we need to cultivate ourselves and others (that is, everyone) as economic subjects with the capacity to innovate and the courage to explore an unmapped terrain.

The final shared concern of the two projects is the goal of bringing a new economy and society into being (though Local/Social Innovation would probably not express themselves in such immodest terms). Focusing on the social economy as a site of innovation and dynamism in contemporary Europe, Local/Social Innovation brings to bear concentrated intellectual resources on an economic sector that has only recently begun to emerge from the shadows. Bringing attention to the sector and its innovative and inclusive potentials is critical to its emergence as a potent social force, helping to attract credibility, resources, and talent to the sector, and to make it the focus of policy initiatives and political agendas. Their project is an example of how attention strengthens that which is attended to, and how social research contributes to consolidati..g realities by making previously marginal sites and activities the focus of widespread interest and deliberate action.

The Community Economies project is more explicit about the role of social research in helping to create the realities it also describes. Whether through presentations and publications or through action research on the ground, they see their work as potentially 'performing' a diverse economy – that is, making it an everyday, commonsense reality by broadening and strengthening the activist imagination, inciting academic investigations among colleagues and students, and suggesting innovative directions for policy intervention. Their vision of the performativity of research foregrounds the ethical decision making of researchers, the choices we make about what to devote attention to and thus to strengthen. The process of ethical self-cultivation that is required to create subjects of community economies is also required of us as academic subjects interested in creating other worlds. This chapter could be seen as a locus and instrument for cultivating a new academic subject, one who regards the academy as an integral and active part of the new worlds being constructed and who seeks academic allies in the process of construction.

References

Community Economies (n.d.), project website <http://www.communityeconomies. org/>.
Gibson-Graham, J.K. (2003), 'An Ethics of the Local', *Rethinking Marxism* 15:1, 49–74.

Gibson-Graham, J.K. (2006), *A Postcapitalist Politics* (Minneapolis, MN: University of Minnesota Press).

Gonzalez, S. and Healey, P. (2005), 'A Sociological Institutionalist Approach to the Study of Innovation in Governance Capacity', *Urban Studies* 42:11, 2055–69.

Ironmonger, D. (1996), 'Counting Outputs, Capital Inputs, and Caring Labor: estimating Gross Household Product', *Feminist Economics* 2:3, 37–64.

Moulaert, F. and Ailenei, O. (2005), 'Social Economy, Third Sector and Solidarity Relations: a conceptual synthesis from history to present', *Urban Studies* 42:11, 2037–53.

Moulaert, F., Delvainquière, J.-C. and Demazière, C. et al. (2000 [2002]), *Globalization and Integrated Area Development in European Cities* (Oxford: Oxford University Press [paperback edn]).

Moulaert, F., Martinelli, F., Swyngedouw, E. and Gonzalez, S. (2005), 'Towards Alternative Model(s) of Local Innovation', *Urban Studies* 42:11, 1969–90.

Moulaert, F. and Nussbaumer, J. (2005), 'The Social Region: beyond the territorial dynamics of the learning economy', *European Urban and Regional Studies* 12:1, 45–64.

Nancy, J.-L. (1992), 'La Comparution/The Compearance: from the existence of "Communism" to the community of "existence"', trans. Strong, T.B., *Political Theory* 20:3, 371–98.

Novy, A. and Leubolt, B. (2005), 'Participatory Budgeting in Porto Alegre: social innovation and the dialectical relationship of state and civil society', *Urban Studies* 42:11, 2023–36.

Social Enterprise Sunderland (n.d.), website <www.socialenterprise-sunderland.org.uk>.

Chapter 3

Microfinance, Capital for Innovation[1]

Mariana Antohi

Microfinance: Definition, Change and Vision

Microfinance refers to a package of financial services (credit, savings, insurance), and in some cases non-financial services (training, follow-up, accompaniment) made available to persons considered poor or excluded from conventional banking services. The current enthusiasm[2] for microfinance operations can be explained by the various significant impacts that such operations are deemed to have on individuals and local communities:

- combating poverty and exclusion (the mechanisms used give priority to those in socio-economic difficulty)
- (re-)injecting some dynamism into local areas by creating economic activity
- supporting projects that respect human factors and the environment (ethical approach).

Microfinance combines sources and experiences from the North and the South. The principle of a loan based on trust is not new and was widely employed in mutual credit practices of the twentieth century[3] (Pierret 2000). The experience

---

1 This chapter was originally written in French. It has been professionally translated, including all quotations from non-English sources.

2 In 1998, the General Assembly of the United Nations (UN) declared 2005 to be the International Year of Microcredit. The aim was to promote microcredit and microfinance throughout the world by highlighting their positive impact on achieving the Millennium development goal to reduce by half the number of people living on less than one dollar per day by 2015. According to the UN, financial exclusion (lack of access to credit) is increasingly seen to be a cause of poverty and other forms of exclusion, rather than simply a consequence (United Nations General Assembly 1997, 1998a, 1998b, 2003a, 2003b). Indeed, over 3 billon members of the active population do not have access to financial services (International Year of Microcredit 2005 n.d.). In 2006, the Nobel Peace Prize was awarded to Muhammad Yunus, founder of the Grameen Bank (Bangladesh), 'the banker of the poor' (Michel 2006).

3 The principle applied in Germany's '*Raiffeisen*' banks spread to other European countries (Netherlands, Switzerland, France) as rural funds or credit houses. European mutualist banks are now fully integrated into the capitalist system and operate according

of the Grameen Bank social credit system in the 1970s in Bangladesh inspired many other programmes and initiatives throughout the world in the 1980s: in Asia, Pacific, Africa, South America, and more recently (in the 1990s) in Europe.

The development of microfinance in Europe has been heterogeneous: in Central and Eastern Europe, microcredit is a dynamic sector,[4] whereas in Western Europe, despite growing interest, the development of microfinance is both more recent and more limited (Underwood and EMN 2006). This development has been influenced by the diversity of national policies and practices, based in different political traditions, socio-historical dynamics and economic contexts. Thus, in Central and Eastern Europe, the microfinance sector is primarily concerned with microcredit for entrepreneurs in an environment marked by inadequate or even non-existent banking offerings after the fall of the Berlin Wall. The situation is very different in Western Europe, where the banking sector is well developed. In this context, microenterprise, supported by microfinance, is seen as a means for encouraging economic growth and social cohesion, helping rebuild local and regional economies where activity has declined (job losses through loss of a local business such as an ironmonger, grocery or garage, or closure or integration/merger of regional companies). In principle, microenterprise helps improve day-to-day life for everyone (not only disadvantaged communities): for example, activities may combat environmental degradation (recovery and recycling), help balance professional and family life (childminding), or respond to the aging population and decline of traditional solutions (homes for the elderly) by providing services in the community (Comeau et al. 2001).

The economic system in Western Europe is characterized by the significant role played by small and medium-sized companies. Each year, approximately 2 million companies are created in the European Union (EU-15), of which 90 per cent are microenterprises with less than five employees (DG Enterprise and Industry 2003). These new companies have both economic and social impacts, and one third of them were created by someone who was previously unemployed.

---

to capitalist principles. However, interest in this model was revived in the South during the 1970s (notably in French-speaking countries), where mutualism was introduced as a development tool based on experience in the North and often implemented with the support of Northern countries (Guérin and Vallat 2000). It was only as of the 1980s that a financial tool created in the South began to spread: since the 1970s, the solidarity credit '*Grameen Bank*' model in Bangladesh has been granting short-term loans, mainly to women, to help them start up an agricultural venture or manual enterprise. This solidarity credit model broke with the principle of prior savings (central to the mutualist credit model) and enabled the very poor to access credit from external sources (lenders, development banks), with concessional terms. The loan is granted without any material or financial collateral and is guaranteed by solidarity within a group to which the beneficiary belongs.

4   According to a joint study conducted by the New Economic Foundation, Evers & Jung, Fondazione Choros and INAISE (2001), the microfinance institutions (MFIs) created in this region after the fall of the Berlin Wall (1989) reached 1.7 million borrowers and 2.3 million savers in only five years, with an annual growth rate of 30 per cent.

The capacity of the banking system to reach and meet the needs of these structures is vital to increasing general socio-economic welfare, but in fact exclusion from banking services constitutes an obstacle to the successful start-up of new revenue-generating activities. Microenterprises frequently encounter problems when looking for finance, whether for working capital or specific investments. In addition to differences in banking traditions (financial instruments used to fund microenterprise) and loan conditions (duration, interest rate, guarantee) between EU member states, the market is far from perfect. Market imperfections (due notably to asymmetrical information) mean that, in some member states, 78 per cent of new companies cannot receive a bank loan (DG Enterprise and Industry 2003, 10). Indeed, they probably constitute the least profitable customers for banks. The poor ratio between transaction costs and profitability/follow-up costs per operation almost systematically disqualifies microenterprises and socioeconomic and solidarity initiatives, which involve too many 'eliminating criteria', from credit for business creation (Lebossé 1998, 19). Microfinance therefore covers a market niche that can offset the reluctance (or refusal) of the banking sector to get involved, providing finance to microenterprises (some of which are socioeconomic initiatives) which meet new needs on new markets[5] (Comeau et al. 2001). Thus microfinance constitutes a 'social response' movement (Lebossé 1998, 24). Although European legislation defines microcredit as corresponding to a maximum amount of 25,000 Euros, minimum loans tend to total between 50 and 5,000 Euros, with an average loan being 12,000 Euros.

Traditionally, microfinance provides a highly standardized financial product – short-term microcredit for creating or developing a specific activity. However, the range of financial services available has become much wider over time, now including savings services and sometimes insurance alongside credit. There has therefore been a series of product innovations as the microfinance customers developed their need for an entire range of products in order to build up their equity, stabilize their expenses and protect themselves against risks. The financial service offering is complemented by a variety of non-financial (such as training, customer support and follow-up) and auxiliary services (sharing good practices, consultation and research into financial operations and measuring social impacts).

Historically a means for combating exclusion from banking services, microfinance institutions (MFI) have based their operating principles on social relations and proximity with the beneficiaries (Lapneau et al. 2004). Numerous microfinance experiences show that taking account of social relations, using values specific to different forms of organization, enables the most vulnerable

---

5   The following new activities are worth mentioning: services to people in difficulty and services at home, new information and communication technologies, assistance and insertion for young people in difficulty, housing improvement schemes, local commerce, tourism, audiovisual technologies, leveraging cultural heritage, local cultural development, waste management, protection and maintenance of nature reserves, energy management, organic agriculture, etc.

customers to be included and hence become economic players in their society and recover their dignity, public recognition and self-confidence. However, as MFIs constitute part of the dominant liberal market system and are subject to pressure from lenders, they have concentrated primarily on ensuring their financial and institutional viability and short-term financial profitability during the consolidation and stabilization period, and have neglected social relations aspects.

The dynamics of granting financial services to those excluded from the traditional banking system is based on balancing *accessibility* and *viability*, which has led to a debate or 'schism' (Morduch 1998) between two approaches to microfinance, the developmental approach and the institutional approach.[6] While some observers see the 'paradigm shift' (Robinson 1995) – or the transition from the developmental vision (based on wide accessibility) to a vision based on financial and institutional viability (increased barriers to entry) – as a 'microfinance revolution' (Robinson 2001), others argue a need to 'put development back into micro-finance' (Fisher and Sriram 2002) or propose 'a middle ground ... that ... will lead to more productive dialogue' between 'two nations divided by a common language' (Woller et al. 1999, 29, 32) or the 'yin and yang' of microfinance (Rhyne 1998).

From the institutional standpoint, the main objective (combating poverty and exclusion) is achieved primarily by ensuring financial self-sufficiency and institutional viability. It is based on financial system development in accordance with a large-scale vision of microfinance whereby profit-making institutions provide high-quality financial services to a large number of clients (*breadth of outreach*) selected from among the least poor of the poor. Institutionalists have drawn up a list of 'best practices' intended to improve the efficiency and effectiveness of microfinance institutions (MFIs) in various domains, such as management, systems management, accounting, marketing, product development. The identification, standardization and large-scale adoption of these best practices is considered an essential step in developing microfinance on an industrial scale as an entry point to capital markets (Sumar 2002). This understanding is rooted in literature developed by the University of Ohio's Rural Finance Programme (US), the World Bank's Consultative Group to Assist the Poor (CGAP), USAID (the US

---

6   Further differentiations are possible in addition to these two approaches: (i) from a poverty reduction standpoint, between *income promotion* and cash management (*consumption smoothing*); (ii) between institutions which emphasize efficiency and effectiveness in reaching their target population (*performance in outreach* and *service delivery*) and those that emphasize the power distribution process (*empowerment*); (iii) between institutions with different legal status and different financial sources: NGOs using donated funds, mutualist institutions and cooperatives using members' own resources, banks and specialized financial institutions; (iv) between different methodologies for granting credit and encouraging saving: MFIs which work through solidarity groups (*solidarity group lending*), those that work through savings and credit systems organized within the community (CBO based MFI), those that give individual credit (*individual lending*) (Coudéré 2004).

Agency for International Development) and academics such as Elisabeth Rhyne and Maria Otero (Otero and Rhyne 1994).

The developmental approach aims specifically to provide the poorest with access to financial services (*depth of outreach*) that have the capacity to enhance the wellbeing of clients and local communities according to a logic of social rather than financial profitability. Although financial self-sufficiency is generally deemed desirable, it does not constitute a baseline condition for the institutional mission. According to this standpoint, applying profit-seeking and market rationale to microfinance distances it from its key social mission, which is the moral cornerstone of and driving force behind the movement (Woller et al. 1999). If lenders grant funds solely on the basis of institutional performance criteria, this will cause some NGOs to 'los[e] their real competitive advantage in the world of development – their capacity to reach the very poorest and engage in a variety of activities that help people change' (Dichter 1996, 259, cited in Woller et al. 1999, 36). In addition, the developmental approach is very critical of the adoption of 'best practices' by MFIs. Such a standard 'blueprint' approach to microfinance is not compatible with the need to innovate and experiment with new products (Woller et al. 1999, 36).

The Social Performance of MFIs

Although financial analysis tools exist to manage the financial performance of MFIs,[7] the social performance of MFIs must also be defined precisely in order to meet the specific challenge facing microfinance, that of reconciling accessibility and viability, social aims and sustainability objectives. This is all the more important given the belief of certain MFIs that reinforcing social performance generates long-term financial benefits.

At present, in addition to ongoing projects to measure the social effects of MFIs,[8] notably through evaluating poverty (CGAP n.d.), two parallel initiatives are attempting to define social performance indicators for MFIs: the SPI initiative[9] implemented by CERISE (n.d.) has been tested and validated, notably by members

---

7   The CGAP (n.d.) defined a series of tools relating to management, operational planning, financial modelling and external auditing (see Isern et al. 2007; Helms and Grace 2004; Lunde 2001; Christen and Rosenberg 1998; Waterfield and Ramsing 1998). The Inter-American Development Bank developed financial performance indicators for improving follow-up and financial visibility (see Von Stauffenberg et al. 2003).

8   Social effects, i.e. changes in clients and non-clients that can be attributed to microfinance, are very difficult to measure as they also constitute the final elements of overall performance (Lapneau et al. 2004).

9   SPI: Social Performance Indicators. The SPI initiative was begun by ethical investors and lenders and implemented by CERISE and its partners in Europe and the South. The CGAP was associated with the steering committee (CERISE n.d.).

of the social solidarity finance unit (FinaSol) of the Workgroup on Solidarity Socio-Economy for a responsible, plural and united world.[10] In 2007, an operational network of microfinance players known as ProsperA was created to promote the culture and practice of social performance by reinforcing the capacities of MFIs and local networks.[11] In Europe, the European Microfinance Platform now aims to place social performance in the microfinance mainstream (European Microfinance Platform n.d.).

According to Lapneau et al. (2004), the performance of an institution, defined according to results obtained at each stage along the intention – action – effects chain, refers to the optimum possibilities of the organization and to the results obtained in terms of defined principles, actions and impacts. Compared with economic and financial performance, the social performance of an organization takes account of the nature of relations between employees and their relations with clients and other partners, as well as of the institution's impact on the 'social conditions of its clients: their standard of living, housing, health, education' (Lapneau et al. 2004, 54). As such, four main dimensions were selected by the SPI initiative to illustrate the social performance of MFIs when considered in their economic and social context (Lapneau et al. 2004, 58):

- Targeting the poor and excluded
- Tailoring products and services to the target population
- Enhancing the social and political capital of clients/empowerment (for example participation in decision-making, beneficiary undertakings to avoid mission slippage, collective action)
- Social responsibility of the MFI: relations with clients and the community.

The way in which these dimensions are integrated into the objectives and practices of various organizations differs according to two types of microfinance (Chao-Beroff et al. 2001, 20):

- *'Pre-banking' microfinance*, the purpose of which is to provide financial services that encourage institutionalization in commercial banking structures to access the capital market and ensure high financial profitability and thus attract private investors;

---

10   The workgroup on solidarity socio-economy for a responsible, plural and united world is a multi-disciplinary debating arena that brings together academics and field players in topic-based workgroups, the aim of which is to promote proposals and strategies for socio-economic transformation (Workgroup on Solidarity Socio-Economy n.d.a).

11   Created in Oaxaca (Mexico) in April 2007 and now also present in Africa, Latin America, Europe and Asia, ProsperA coordinates exchanges and joint actions based on the SPI tools and the initiatives of its members, notably in terms of governance, impact and social performance appraisal (Workgroup on Solidarity Socio-Economy n.d.b).

- *Social solidarity finance*, which is defined according to several factors:
  a) *mission*: using a financial tool to promote fair and sustainable development;
  b) *long-term vision*: increasing social capital;
  c) *specific identity*: distinguishing it from a multitude of players with different techniques and behaviour patterns;
  d) *competencies*: thinking globally and federating individual players around financial activities, recognizing the needs of individual entrepreneurs and communities according to their social and economic conditions;
  e) *business*: financing activities and persons within the context of public interest;
  f) env*ironment*: poverty, exclusion and limited access to financial services.

Some analysts correlate the current crisis in the microfinance sector – illustrated by phenomena such as mass loss of clients, falling transaction volumes or un-recovered loans – with the fact that certain MFIs have ceased to take account of social relations in their search for financial profitability (Quiñones and Sunimal 2003, 3).

Microfinance for Social Solidarity: Innovations

Solidarity microfinance is socially innovative in several respects, as identified by Moulaert (2000) and Hillier et al. (2004):

- *the 'content' or 'final purpose' dimension* (in terms of socio-economics and solidarity): meeting needs that are not met through market mechanisms, granting the right to receive credit and take economic initiatives to project holders who are excluded from the conventional banking system;
- *the 'process' dimension*: the dynamics which mobilize both 'solidarity' and 'participation' require changes in social relations between individuals and groups which 'should enable their needs to be met and enhance the participation of excluded groups in decision-making' (Hillier et al. 2004, 135);
- *the 'empowerment' and 'agency' dimension*: seeking to enhance the social and political capital of clients, thereby 'increasing socio-political capacity and access to the resources needed to meet human needs and increase participation' (Hillier et al. 2004, 135).

Tailoring Methods and Products to Needs

In attempting to combat exclusion through integration, solidarity microfinance helps build, develop and/or reinforce social relations and social cohesion.

However, this new and innovative interaction in favour of inclusion (Moulaert et al. 2005) occurs via 'changes in programmes and institutions conducive to the inclusion of excluded individuals and groups' (Hillier et al. 2004, 150). The capacity of microfinance to meet the credit needs of project holders depends on its 'capacity to organize social relations through institutional forms that illustrate and encourage values of reciprocity and solidarity and enable those needs to be addressed' (Hillier et al. 2004, 141). It is here that the solidarity standpoint differs from the banking standpoint, with its methods of interaction with and acknowledgement of social relations, and their advantages and costs for MFIs. In particular, these methods refer to the following:

- *Solidarity and participation*, the implications of which characterize the operating principle of social guarantee groupings, cooperative systems where all member-beneficiaries help manage the institution, village-based banks where an entire village is involved in and responsible for managing the funds correctly for the benefit of the village, etc. (Lapneau et al. 2004, 51).
- *Services based on proximity with beneficiaries* (Lapneau et al. 2004, 52):
  a) geographical proximity with rural development agencies or 'mobile banking' services that reach out to clients;
  b) social proximity in seeking to reduce barriers between clients and the institution (local agents, services specific to the cultural or religious context, etc.);
  c) temporal proximity with frequent contact between the institution and its clients, who make regular repayments or benefit from frequent training and exchange sessions. Proximity enhances trust, reduces information imbalances and attenuates social barriers between clients and the institution (Servet 1996, cited in Lapneau et al. 2004, 52).
- *Services for the excluded population*, designed and *tailored* to the needs of an economically and socially marginalized population. When talking about access of excluded persons to financial services, the term *access* can mean both 'the possibility of owning something', and 'something that can be understood', that is, is intelligible (Larousse 1992, 18). Exclusion therefore has a double meaning in terms of lack of access which, when applied to banking products and services, means 'both not having these products available and, if the products are available, not being able to use them in an appropriate manner' (Gloukoviezoff 2002, 216).
- Accessibility can be extended via a client-centred strategy which specifically addresses the obstacles facing the poor and the instruments that may help them. According to Von Pischke (1991), this approach requires some innovation in terms of funding, to provide the services and instruments either at least-possible cost or with a higher performance level, or both. Three forms of *funding innovation* were identified (Von Pischke 1991, 366):

a) extending payment schedules on financial markets
b) reducing transaction costs
c) enhancing the evaluation process within MFIs by creating new strategies that mobilize an innovative vision of evaluation issues.

In addition to these technical forms of innovation, *non-financial training, customer support and follow-up services* are also provided to business creators. Numerous studies (Gloukoviezoff 2005; Guérin and Servet 2003; Servet and Guérin 2002; Servet and Vallat 2001; Servet 1999; Labie 2000; Hulme and Moseley 1996) have shown that financial services, although a key point, are not the only difficulty facing business creators. Many problems must be addressed when creating and developing microenterprises, including a lack of training, lack of information, problems with procuring raw materials, marketing and interpreting the legislative framework (Labie 2000). Long-term assistance and customized, multi-dimensional follow-up therefore are considered by this approach to be 'key success factors' (Comeau et al. 2001, 109). Indeed, long-term assistance should cover the pre-start-up and start-up phases, as well as the business ramp-up. Customized, multidimensional support, based on analysing the plans and personal situation of the founder, reduces both technical risks (mitigated by, for instance, project design, response to administrative requirements such as registration and legislation, development of sustainable competencies and promotion) and psycho-social risks (mitigated by group support, seminars, documentation, individual meetings). These nonfinancial services have a significant impact on the *human capital* available to the project holder, defined according to the latter's 'knowledge, qualifications, competencies and personal qualities available to facilitate the creation of personal, social and economic well-being' (OCDE 2001, 18).

The methods and products disseminated by MFIs may reinforce social relations (which create social capital), or undermine or break them, given that the quality of social relations indicates the extent of social capital. 'Social capital' can be defined as the capacity of individuals to cooperate and act together, using or creating social links to achieve common goals for mutual benefit (Montgomery 1997, cited in Quiñones and Sunimal 2003, 3). Social capital[12] is a manifestation of the interactions between values shared by individuals and the institutions and structures created to reconcile these values. These values may be subordinate to the desires and interests of each individual, but they are nevertheless shared and determine collection action (Woolcock and Narayan 2000).

---

12 The SPI initiative identified the following indicators to measure the level of social capital: MFI client participation in decision-making, allocation of resources, client capacity to analyse situations and formulate projects, their capacity to organize and manage themselves, establish relations with administrative and political authorities, negotiate with technical managers, take control of the financial system implemented, identify good and bad measures and decisions, address the longer term, and integrate tools regarding territorial dynamics.

The rich literature on social capital shows that organizations which take account of existing solidarity and social relations and seek to reinforce them through their intervention methods establish a more trusting relationship with their target population and have a more positive impact on the living conditions of their clients.[13] This is achieved through products that are designed and adapted to the specific needs of the latter (women, young people, ethnic minorities, unemployed persons and so on) in their context (urban, rural), with consideration given to group size, diversity of activity among members of the group, physical proximity between them, leadership quality, and the type of non-financial services proposed by the institution. In this respect, methods that favour training and information reinforce trust and create social capital. As shown in several case studies (Gloukoviezoff 2005; Guérin and Servet 2003; Servet and Guérin 2002; Servet and Vallat 2001; Servet 1999) and theoretical analyses (Chao-Béroff et al. 2001, 30–31), the products proposed by MFIs may:

- Help create/strengthen social capital, notably:
  a) voluntary, flexible, accessible and secure savings programmes;
  b) the possibility of independently defining the amount, duration and payment dates for credit repayments;
  c) insurance products (for example life insurance, health insurance) considered by clients to reinforce social relations (as family members no longer have to bear the cost of unpaid loans);
  d) the possibility of subscribing to a common guarantee fund at reduced cost.
- Undermine social relations, as in the following cases:
  a) regular compulsory savings payments or those debited directly;
  b) weekly, increasing credit repayments;
  c) group funds which work to rules not defined by members of that group.

Social Relations and Social Capital, Long-term Gains

The capacity of microfinance to constitute an innovative, inclusive intervention must be correlated with the use of methods and products that help create/strengthen social relations and social capital. This implies:

- knowledge of existing social relations on the basis of which social capital can be reinforced;
- methods and products designed to stimulate social relations and social capital, and so tailored to the target population;

---

13   Unger (1998) states that, in communities where trust between individuals is limited, the number of cooperative organizations is also limited; Fukuyama (1995) notes that trust enables individuals to work on private projects in a cooperative way.

- provision for monitoring the effects of the intervention on existing social relations and those that it helps create.

From the banking standpoint on microfinance, the costs incurred by MFIs in addressing these issues are not compatible with objectives of efficiency and financial profitability. MFIs that practise social finance confirm that reinforcing social relations constitutes a long-term gain for their institution, the challenge being not to succumb to market pressure for immediate financial profitability.

The debate concerning the financial profitability of MFIs relates to a wider debate concerning the field of application of microfinance, that is, the domain of social economics and solidarity, where the non-monetary logic of empowerment and social relations takes priority over that of profit maximization. The *social utility* or social profitability dimension relates to the social and professional integration of disadvantaged populations and the production of goods and services not provided by the state. These goods and services boost the local economy, create social links, establish trust between local players, and improve the quality of life and general well-being of the population (Bouchard et al. 2003). Such externalities imply an 'interdependence between agents that cannot be represented by market price mechanisms' (Bouchard et al. 2003, 9) and which challenges the 'dichotomy between public and private goods and enhances the role of collective action in economic effectiveness' (Aglietta 1998, cited in Bouchard et al. 2003, 9). As the specific purpose of such actions is to reintroduce a social element into economics, they cannot be appraised according to monetary valuation techniques which would require them to justify their existence according to economic criteria.

For microfinance to be viable, it must meet challenges of efficiency, profitability and sustainability, but the definition of these terms varies according to the 'banking' or 'solidarity' standpoint.

From the 'banking' standpoint, *efficiency* is defined in terms of repayment quality and institutional efficiency, notably as regards the capacity of an institution to achieve 'a regularly increasing volume of credit and savings, which has significant weight in the economy', ensure 'repayment rates of close to 100%', and 'transform itself into a permanent, efficiently-managed, financial institution that achieves organizational and financial autonomy in a medium-term time frame (5 to 12 years)' (IRAM 1998, 91). From the 'solidarity' standpoint, the efficiency of an institution is defined according to its 'capacity to provide tailored products and services that have a positive impact on beneficiaries' (Chao-Béroff et al. 2001, 33). Efficiency is thus rooted in suitable methods and products, and results from proximity with clients, service quality and services that correspond to needs. The efficiency of solidarity finance is specifically defined according to existing social situations. Taking account of social relations and social capital generates value-added both for the institution (for example reduced transaction costs) and for the client: reduced distance between the institution and the cultural background of its poorest clients, increased information, enhanced competencies developed through exchange, improved participation (Chao-Béroff et al. 2001).

The *profitability* of an MFI is defined in terms of costs and resources. Solidarity finance practices, which mobilize and reinforce social relations, incur high initial costs relating to training and developing groups, personal development, training, monitoring and appraising social relations, and studying social relations (time devoted to each client). Building social links and social capital between an institution and its clients is a slow process, so these costs only decrease in the long term (after several years) when gains will become visible in terms of improved portfolio quality, client loyalty, client participation in monitoring activities, repayment discipline, and greater credit agent productivity because of training received by clients and groups. If short-term financial profitability is sought, this implies certain risks for the institution and its clients: exclusion of the poorest clients and marginal regions, reduced social commitment, loss of experience and know-how as regards working with the most disadvantaged, and undermined long-term viability of the institution and of its clients and groups (Chao-Béroff et al. 2001, 34).

Although acknowledgement of social relations incurs significant costs, it generates long-term efficiency and stability, and hence enhances the institutional *sustainability* of microfinance, reducing vulnerability to sudden shocks and tailoring its activity to the intervention context. Chao-Béroff et al. (2001) therefore claim that the debate between accessibility and viability is a red herring. The real question is not whether or not microfinance should take account of social relations, but who should bear the additional costs generated by acknowledging them. Recurrent costs may be covered by clients or groups;[14] that is, given the public interest nature of social relations and social capital, by public or private lenders.

Solidarity finance seeks to integrate social relations issues into interventions in order to better meet needs. It uses financial capital not in the market sense of 'investment of resources with expected returns in the marketplace' (Lin 2001, 3), but in terms of social purpose, acting on human and social capital and ultimately enhancing inclusion.

This is an innovation 'in terms of social relations' which 'reinforces the means of inclusion, and therefore reinforces social capital at the local level' (Hillier et al. 2004, 142).

By making financial capital accessible to those excluded from the conventional financial market, microfinance opens up a new space for transactions. It facilitates *access* to this new space, and this is possible only by *building/reproducing social capital*. Solidarity microfinance redefines the notion of capital by restructuring relations between different types of capital at the local level: financial capital, human capital, social capital (Figure 3.1).

Interactions between these different types of capital, and their capacity for renewal, relate to issues of local development and new local governance relations

---

14   For example, by including these costs when calculating the interest rate, through voluntary client participation in training and management activities, and so on.

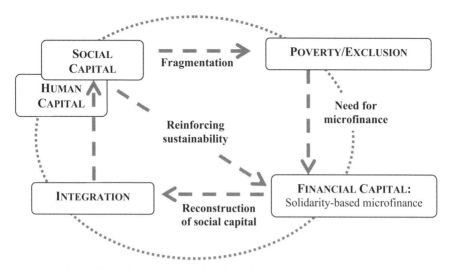

**Figure 3.1   Microfinance: Mobilisation of financial capital
for the (re)construction of social capital**

which, 'instead of relying on a single type of agent (private, public), depend on the
capacity of several agents to work together' (Hillier et al. 2004, 142).

New Challenges: Developing Synergies and Complementarities
that will overcome the Fragmented Vision of Social Exclusion

Microfinance claims to be an important tool for combating poverty and exclusion in
Europe. Several studies and reports (Microfinance Centre for Eastern Europe and the
New Independent States 2007) have addressed issues relating to this position: issues
of financial exclusion, key obstacles, challenges and political recommendations
for supporting the sector in both Western and Eastern Europe (Evers and Lahn
2006), benchmarking and social performance, and promoting innovative products
and approaches (for example the BDS – Business Development Support, technical
assistance, developing an innovative product offering and so forth).

In the current context, combating poverty and social exclusion involves a series
of new challenges that are not addressed, or inadequately addressed, by the ideas and
practices of microfinance. There is a need to go beyond the fragmented response to
financial aspects of exclusion, to identify and test synergies and complementarities
with more conventional responses to poverty and social exclusion (social policy,
developmental NGOs, and so on) and with ethical and solidarity-based citizen
initiatives (solidarity savings plans, ethical investment products, and so forth), in
order to address the issue of poverty and social exclusion in all its complexity.

Exclusion has been an issue for several decades, and changes in analysis over this time have resulted in changed social policies for combating poverty and exclusion. However, given the concurrent changes in poverty and social exclusion (Moulaert et al. 2007; de Haan 1999; Castel 1995; Touraine 1991, 1992), social policy, as an expression of collective responsibility through the state, must respond to the challenges raised by a new context marked by increased global competition, an ageing population and declining birth rate, new technologies, high unemployment, new forms of poverty, and so on. Whereas there seems to be some agreement that multiple processes are responsible for exclusion – requiring some explanation of their inter-relations – and that multiple spatial and institutional factors also play a role (Moulaert et al. 2007), the social policies defined and implemented as a response are often fragmented, targeting one specific category of the poor or excluded. The results are often assessed according to quantitative indicators that are designed to meet the needs of administrative management (number of persons helped, number of days of assistance provided, amount of aid per person, and so on) and do not give sufficient attention to the dynamics which enable excluded persons to reintegrate with their society, that is rebuilding their social relations and social capital.

In parallel, civil society organizations have developed which are specialized in serving particular groups of the poor (those excluded from the conventional financial market, the highly indebted, homeless, immigrants and ethnic minorities, Roma, handicapped persons, children, and so on).[15] The terms of collective responsibility have been altered by changing ideas regarding public assistance.[16] For example, the complexity of issues can be highlighted by current debates on poverty and exclusion and on *individual responsibility*, *collective responsibility* and their inter-relations.

Given the 'great difficulty in discerning the difference between mere political grandstanding and positions that are genuinely significant in terms of concrete achievement' (Editorial 2004, 8), European microfinance players have realized the importance of creating structures at all levels (local/regional, national and European) to help develop synergies and a unified voice for engaging in ongoing political dialogue with the public authorities. Over several years, several major

---

15   Within the EU, and in the general context of the Method of Open Coordination on Social Protection and Social Inclusion and National Action Plans to combat poverty and social exclusion instituted by the Lisbon strategy (2000), several networks of civil society players act as structural advisors for the European Commission's social policy. These include the European Microfinance Network, Eurochild, European Anti-poverty Network (EAPN), European Federation of National Organisations Working with the Homeless (FEANTSA) (Directorate-General for Employment, Social Affairs and Equal Opportunities n.d.a).

16   A new concept of *active inclusion* has emerged: mobilization of the poor and excluded in order to achieve their insertion into the labour market while leveraging available assistance and support (Directorate-General for Employment, Social Affairs and Equal Opportunities n.d.b).

microfinance players (including MFC – Microfinance Centre for Central and Eastern Europe and the New Independent States; REM – the European Microfinance Network), ethical and solidarity finance players (FEBEA – the European Federation of Ethical and Alternative Banks and Financiers) and citizen-based ethical and solidarity initiatives[17] have emerged to address issues of microfinance and explore the possibilities for solidarity finance and new forms of economic solidarity and responsibility that European society has opened up in response to the challenges of poverty and social and environmental exclusion.

The capacity of microfinance, or ethical and solidarity finance in general, to comprise innovative and integrated responses is particularly relevant to two of the challenges (European Inter-Network of Responsible Economy Initiatives 2007) currently facing those engaged against poverty and social exclusion:

- the need to go beyond a fragmented, palliative view of problems of poverty and financial exclusion, and promote synergies between approaches and services so that poverty and exclusion can be addressed in their full complexity;
- the need to address the question of individual commitment in the fight against exclusion and poverty, notably through ethical market choices such as solidarity savings, ethical investment and responsible consumption. These forms of expressing citizenship concerns through the economy, in addition to being vectors of mobilization, raise new possibilities for combating poverty and social exclusion. They refer to a development model in which social and environmental issues concerning living and working opportunities for all, access to a good quality Welfare State, active participation in society choices, and the availability and renewal of natural resources constitute key structural elements (Oliveri 2005).

There are several important points raised by these two challenges:

- the need to explore and *identify complementarities* between the specific microfinance approach and more conventional forms of combating poverty and exclusion (as practised by public authorities and developmental NGOs, and so on), as well as with citizen-based initiatives;

---

17 The European Inter-Network of Responsible Economy Initiatives (IRIS), created in 2007 as part of the dialogue platform for ethical and solidarity-based initiatives (Council of Europe), forms a forum for dialogue and cooperation between European and International networks representing the various types of citizen-based responsible economy initiatives: ethical and solidarity finance (FEBEA, INAISE), fair trade (IFAT), responsible consumption (URGENCI, ASECO), socio-economic insertion based on solidarity (ENSIE) with participation and support from institutional partners (Council of Europe, the Autonomous Province of Trento) (European Inter-Network of Responsible Economy Initiatives n.d.).

- the need to create a *conceptual framework* to highlight these complementarities and the value that they add;
- the need to *try out new means of expressing these complementarities*, notably using financial instruments tested in the social innovation domain.

The Multipartite Social Contract (MSC)[18] pilot project constitutes an interesting example of leveraging complementarities between various approaches to combating poverty and integrating excluded persons. Although it is not a microfinance project, it is an integrated approach that addresses the multiple dimensions of poverty and exclusion and includes citizen-based strategies such as responsible consumption and solidarity finance as factors which reinforce the efficiency and effectiveness of these strategies.

The MSC addresses various categories of social problem[19] (over-indebtedness, long-term unemployment, excluded young people and so on) by mobilizing a wide range of players (public authorities, NGOs specializing in social work, citizens). The aim is to integrate the various dimensions of work against exclusion – social support and follow-up, access to employment, financial resources, social contacts and so on – by mobilizing organizations specialized in one or more of these domains alongside the beneficiaries themselves. Various forms of support and solidarity are mobilized in this way: financial support through solidarity finance, principles of responsible consumption, access to certain basic products with a social solidarity perspective, conventional support and follow-up. This approach helps to integrate the social functions of each player, enabling each to play a key role in creating a more equitable society and improving understanding of links between consumption choices and quality jobs or environmental benefits. The complementarities, responsibilities and commitments of each player are defined by mutual agreement, and a key innovative element is the participation of beneficiaries as partners in defining the inclusion roadmap. Integrating approaches makes it easier to express complementarities between forms of support and solidarity which, in addition to being mutually enriching, help build synergies (and therefore improve efficiency).

---

18   The Multipartite Social Contract (MSC), created by the Council of Europe as part of the dialogue platform for ethical and solidarity-based initiatives, is a European pilot project launched in 2007 by the Inter-Network of Responsible Economy Initiatives (IRIS) and funded jointly by the Council of Europe and the European Commission within the framework of the Community Action Programme against social exclusion 2002–2006 (PROGRESS) and coordinated by the Regional Chamber for Social Debt (CRESUS Alsace). For further information regarding this tool, see Council of Europe website (n.d.).

19   The experimental phase of the Multipartite Social Contract (February–December 2007) targeted fifteen highly indebted residents of Strasbourg (France) who became 'partner beneficiaries' rather than simply 'beneficiaries'. In December 2007, the MSC was extended to the long-term unemployed whose benefit entitlement had expired.

Conclusions

Created as a means for enhancing the inclusion of those excluded from the banking system and mainstream society, MFIs initially based their operations on principles of social relations and proximity with beneficiaries. However, these principles were often discarded during the consolidation and stabilization period under pressure from lenders and the dominant neoliberal system. As such, the 'solidarity' standpoint of the initial microfinance mission now differs from the 'banking' standpoint, notably because of the importance the former attributes to social relations and social capital, despite the significant short-term costs generated by this. The 'gains' for the institution, clients and community are visible in the longer term.

Solidarity microfinance is socially innovating as regards several of the dimensions defined by Moulaert (2000; Hillier et al. 2004): the 'content' or 'purpose' dimension (meeting unmet needs), the process dimension process (a dynamic which mobilizes 'solidarity', 'participation', 'reciprocity'), and the empowerment dimension (improving the client's social and political capital).

Solidarity microfinance constitutes a new and innovative means of inclusive interaction, as it integrates individuals and groups through products and methods that are tailored to requirements and take account of social relations in order to better meet the needs identified. Financial capital is therefore used to achieve the social purpose of integration, and also improve human and social capital: it helps combine 'social' and 'economic' concerns.

In the current context, strategies for combating poverty and social exclusion raise new challenges that require responsibilities (individual and collective) to be redefined and the multiple dimensions of poverty and/or exclusion to be considered in full. These issues require microfinance to look beyond its particular specialization (combating financial exclusion by providing access to financial and non-financial services that help stimulate economic activity) towards repairing the problems of exclusion. In particular, this implies identifying and testing synergies and complementarities with more conventional approaches to combating poverty (public authorities, social service NGOs and so on) and promoting new responsibilities and commitments (citizen-based ethical and solidarity initiatives such as solidarity savings products and responsible investment). This would enable the microfinance movement to return to its original *raison-d'être*:

> The micro-credit movement which is built around and for and with money, ironically, is at its heart, at its deepest root, not about money at all. It is about helping each person achieve his or her fullest potential. It is not about cash capital, but about human capital. Money is merely a tool that helps unlock human dreams and helps even the poorest and the most unfortunate people on this planet achieve dignity, respect and meaning in their lives. (Yunus and Jolis 1998, 279)

References

Affichard, J. and de Foucauld, J.B. (eds) (1992), *Justice sociale et inégalité* (Paris: Esprit).

Aglietta, M. (1998), *Accumulation et crises du capitalisme* (Paris: Odile Jacob).

Blanc, J., Guérin, I. and Vallat, D. (eds) (2000), *Exclusion et liens financiers, Rapport du Centre Walras 1999–2000* (Paris: Économica).

Bouchard, M.J., Fontan, J.-M., Lachance, E. and Fraisse, L. (2003), 'L'évaluation de l'économie sociale, quelques enjeux de conceptualisation et de méthodologie', *Cahiers du CRISES*, International Collection, March, IN0301.

Castel, R. (1995), *Les métamorphoses de la question sociale. Une chronique du salariat* (Paris: Fayard).

CERISE (n.d.), *Comité d'Echanges, de Réflexion et d'information sure les Systèmes d'Epargnes crédit* [website], <http://www.cerise-microfinance.org>.

CGAP (n.d.), 'Assessing the Relative Poverty of Microfinance Clients: a CGAP operational tool', *CGAP* (published online) <http://www.cgap.org/portal/binary/com.epicentric. contentmanagement.servlet.ContentDeliveryServlet/Documents/TechnicalTool_05_overview.pdf<http://www.cgap.org/portal/binary/com.epicentric.contentmanagement.servlet.ContentDeliveryServlet/Documents/TechnicalTool_05_overview.pdf>.

Chao-Béroff, R., Prébois, A. and Iserte, M. (2001), 'Une finance solidaire pour retisser les liens sociaux', *Cahiers des propositions pour le XXIe siècle*, Workgroup on solidarity socio-economy for a responsible, plural and united world, 17, 1–78 (Paris: Charles Léopold Mayer).

Christen, R.P. and Rosenberg, R. (1998), 'External Audits of Microfinance Institutions', *Technical Tool Series* No. 3, Vol. 1, December 1998 (Washington, DC: CGAP/World Bank) (published online) <http://collab2.cgap.org/gm/document-1.9.2999/TechnicalTool_03_v1.pdf>.

Comeau, Y., Favreau, L., Lévesque, B. and Mendell, M. (2001), *Emploi, économie sociale, développement local: les nouvelles filières* (Sainte-Foye: Presses de l'Université du Québec).

Coudéré, H. (2004), 'Micro-Finance: Une introduction', *Interfaces* 4:23, 2–5.

Council of Europe (n.d.), [website] <http://www.coe.int>.

de Haan, A. (1999), 'Social Exclusion: towards an holistic understanding of deprivation', paper presented at the *World Development Report Forum, Inclusion, Justice and Poverty Reduction: a dialogue* (Berlin, February 1999).

Dichter, T. (1996), 'Questioning the Future of NGOs in Microfinance', *Journal of International Development* 8:2, 259–69.

Directorate-General for Employment, Social Affairs and Equal Opportunities (n.d.a), *Social Inclusion* [webpage] <http://ec.europa.eu/employment_social/spsi/poverty_social_exclusion_ en.htm>.

Directorate-General for Employment, Social Affairs and Equal Opportunities (n.d.b), *Active Inclusion* [webpage], <http://ec.europa.eu/employment_social/spsi/active_inclusion_en.htm>.

Directorate-General for Enterprise and Industry (2003), *Le microcrédit pour la petite entreprise et sa création: combler une lacune du marché*, D 324 10/ n°14261.

Editorial (2004), 'A la recherche de l'économie solidaire', *Finance&Common Good/Bien commun* 20, 5–9.

European Inter-Network of Responsible Economy Initiatives (n.d.), [website] <http://www.iris-network.eu>.

European Inter-Network of Responsible Economy Initiatives (2007), *Guide à réflexion méthodologique pour une large implication des citoyens* (Preliminary draft for discussion), created for the Council of Europe's dialogue platform for ethical and solidarity-based initiatives against poverty and social exclusion (DG III – Development of social cohesion) and the European Union's Community Action Programme against social exclusion 2002–2006 (PROGRESS).

European Microfinance Platform (n.d.), *European Microfinance Platform: Networking with the South* [website], <http://www.microfinance-platform. eu/>.

Evers, J. and Lahn, S. (2006), 'Promoting Microfinance, Policy Measures Needed', *Finance&Common Good/Bien commun* 25, 47–53.

Fisher, T. and Sriram, M.S. (eds) (2002), *Beyond Micro-Credit: putting development back into micro-finance* (Oxford: Oxfam and New Economics Foundation).

Fukuyama, F. (1995), *Trust: the social virtues and the creation of prosperity* (New York: The Free Press).

Geyer, E. (ed.) (2007), *International Handbook of Urban Policy, Volume 1: Contentious Global Issues* (Cheltenham: Edward Elgar).

Gloukouviezoff, G. (2002), 'Des pratiques bancaires sources d'exclusion', in Servet and Guérin (eds).

Gloukoviezoff, G. (ed.) (2005), *Exclusion et liens financiers. Rapport du Centre Walras 2004* (Paris: Économica).

Guérin, I. and Servet, J.-M. (eds) (2003), *Exclusion et liens financiers. Rapport du Centre Walras 2003* (Paris: Économica).

Guérin, I. and Vallat, D. (2000), 'Eléments de réflexion critique : introduction', in Blanc et al. (eds).

Helms, B. and Grace, L. (2004), 'Microfinance Product Costing Tool', *Technical Tools Series* No. 6, June 2004 (Washington, DC: CGAP/World Bank) (published online) <http://collab2.cgap.org/gm/document-1.9.3005/TechnicalTool_06.pdf>.

Hillier, J., Moulaert, F. and Nussbaumer, J. (2004), 'Trois essais sur le rôle de l'innovation sociale dans le développement spatial', *Géographie, Economie, Société* 6:2, 129–52.

Hulme, D. and Moseley, P. (1996), *Finance against Poverty*, vols I and II (London: Routledge).

International Year of Microcredit 2005 (n.d.), [project website] <http://www.yearofmicrocredit.org/>.

IRAM (1998), 'Orientation sur la préparation et l'exécution des interventions' (publishedonline)<http://europa.eu.int/comm/development/body/publications/microfinance/p026_fr.pdf>.

Isern, J., Abrams, J. and Brown, M. (2007), *Appraisal Guide for Microfinance Institutions, Resource Manual* (Washington DC: CGAP/World Bank) (published online) <http://collab2.cgap.org/gm/document-1.9.2972/MFIResourceGuide.pdf>.

Labie, M. (2000), *La microfinance en questions. Limites et choix organisationnels* (Brussels: Luc Pire).

Lapneau, C., Zeller, M., Greely, M., Chao-Béroff, R. and Verhagen, K. (2004), 'Performances sociales, une raison d'être des institutions de microfinance et pourtant encore peu mesurées. Quelques pistes', *Monde en développement* 2:126, 51–68 (published online) <http://www.cairn.info/revue-mondes-en-developpement-2004-2-page-51.htm>.

Larousse (1992), *Grand Larousse* vol. 1 (Paris: Larousse).

Lebossé, J. (1998), *Microfinance et développement local*, Montréal: Institut de formation en développement économique communautaire (IFDEC) (Paris: OCDE).

Lin, N. (2001), *Social Capital: a theory of social structure and action* (Cambridge: Cambridge University Press).

Lunde, S.A. (2001), 'Using Microfin 3: a handbook for operational planning and financial modeling', *Technical Tool Series* No. 2, September 2001 (Washington, DC: CGAP/World Bank) (published online) <http://www.microfinancegateway.org/files/3246_TechnicalTool_02.pdf>.

Michel, A. (2006), 'Muhammad Yunus, un Nobel "prêteur d'espoir"', *Le Monde*, 15 October 2006 (published online) <http://www.lemonde.fr/international/article/2006/10/14/muhammad-yunus-un-nobel-preteur-d-espoir_823502_3210.html>.

Microfinance Centre for Central and Eastern Europe and the New Independent States, European Microfinance Network and Community Development Finance Association (2007), *From Exclusion to Inclusion through Microfinance*, study financed by the EC, DG for Employment, social affairs and equal opportunities (published online) <http://www.mfc.org>.

Montgomery, J.D. (1997), 'Are Asian Values Different ?', in Montgomery, J.D. (ed.), *Values in Education: social capital formation in Asia and Pacific* (Hollis, NH: Hollis Publishing).

Morduch, J. (1998), 'The Microfinance Schism', Development Discussion Paper 626, 1–15, February 1998, Harvard Institute for International Development (published online) <http://www.cid.harvard.edu/hiid/626.pdf>.

Moulaert, F. (2000), *Globalization and Integrated Area Development in European Cities* (Oxford: Oxford University Press).

Moulaert, F., Martinelli, F., Swyngedouw, E. and Gonzales, S. (2005), 'Towards a Conceptualization of Social Innovation in Community Development', *Urban Studies* 42:11, 1969–90.

Moulaert, F., Morlicchio, E. and Cavola, L. (2007), 'Social Exclusion and Urban Policy in European Cities: Combining "Northern" and "Southern" European Perspectives', in Geyer (ed.).

New Economic Foundation, Evers & Jung, Fondazione Choros and INAISE (2001), *Finance for Local Development: new solutions for public-private action*, study financed by the EC, DG Employment and Social Affairs (published online) <www.localdeveurope.org>.

OCDE (2001), *Du bien-être des nations: le rôle du capital humain et social* (Paris: OCDE).

Oliveri, F. (2005), 'Le potentiel effectif de développement des solidarités dans le marché: synthèse des débats du Forum 2004', *Tendances de la cohésion sociale*, Editions du Conseil de l'Europe 14, 219–46.

Otero, M. and Rhyne, E. (1994), *The New World of Microentreprise Finance: building health institutions for the poor* (West Hartford, CT: Kumarian Press).

Pierret, D. (2000), 'Programmes de microcrédit du Nord et du Sud. Croisement des sources d'inspiration et cloisonnement des réflexions', in Blanc et al. (eds).

Quiñones, B. and Sunimal, F. (2003), 'Le capital social dans la finance sociale', *Chantier Finance solidaire*, Workgroup on Solidarity Socio-Economy for a responsible, plural and united world (published online) <www.socioeco.org>.

Rhyne, E. (1998), 'The Yin and Yang of Microfinance: reaching the poor and sustainability', *MicroBanking Bulletin* 2:1, 6–8.

Robinson, M. (1995), 'The Paradigm Shift in Microfinance: a perspective from HIID', Development Discussion Paper 510, 1–36, Harvard Institute for International Development, May 1995 (published online) <http://www.cid. harvard.edu/hiid/510.pdf>.

Robinson, M. (2001), *The Microfinance Revolution, Sustainable Finance for the Poor* (Washington DC: The World Bank).

Servet, J.-M. (1996), 'Risque, incertitude et financement de proximité en Afrique. Une approche socio-économique', *Revue Tiers-Monde*, 37:145, 41–58.

Servet, J.-M. (ed.) (1999), *Exclusion et liens financiers. Rapport du Centre Walras 1999–2000* (Paris: Économica).

Servet, J.-M. and Guérin, I. (eds) (2002), *Exclusion et liens financiers. Rapport du Centre Walras 2002* (Paris: Économica).

Servet, J.-M. and Vallat, D. (eds) (2001), *Exclusion et liens financiers. Rapport du Centre Walras 2000–2001* (Paris: Économica).

Sumar, N. (2002), *Poverty Alleviation through Credit: catapulting microfinance to Wall Street* (Ontario: Queen's School of Business).

Touraine, A. (1991), 'Face à l'exclusion', *AAVV Citoyenneté et urbanité* (Paris: Esprit).

Touraine, A. (1992), 'Inégalités de la société industrielle, exclusion du marché', in J. Affichard and de Foucauld (eds).

Underwood, T. and the European Microfinance Network (2006), 'Overview of the Microcredit Sector in Europe 2004–2005', EMN Working Paper 4 (Paris: EMN).

Unger, D. (1998), *Building Social Capital in Thailand* (Cambridge: Cambridge University Press).

United Nations General Assembly (1997), *Role of microcredit in the eradication of poverty*, Resolution A/RES/52/194 adopted by the General Assembly on the Report of the Second Committee (A/52/628/Add. 6), fifty-second session of the General Assembly, Agenda item 97(*f*) (published online 18 December 1997) <http://www.undemocracy.com/A-RES-52-194.pdf>.

United Nations General Assembly (1998a), *Role of Microcredit in the Eradication of Poverty*, Report of the Secretary-General A/53/223, fifty-third session of the General Assembly, item 101 of the Provisional Agenda: Implementation of the First United Nations Decade for the Eradication of Poverty (1997–2006) (published online 10 August 1998) <http://www.undemocracy.com/A-53-223.pdf>.

United Nations General Assembly (1998b), *International Year of Microcredit, 2005*, Resolution A/RES/53/197 adopted by the General Assembly on the Report of the Second Committee (A/53/613), fifty-third session of the General Assembly, Agenda item 98 (published online 15 December 1998) <http://www.undemocracy.com/A-RES-53-197.pdf>.

United Nations General Assembly (2003a) *Programme of Action for the International Year of Microcredit, 2005*, Resolution A/RES/58/221 adopted by the General Assembly on the Report of the Second Committee (A/58/488), fifty-eighth session of the General Assembly, Agenda item 98 (published online 23 December 2003) <http://www.undemocracy.com/A-RES-58-221.pdf>.

United Nations General Assembly (2003b), *Implementation of the first United Nations Decade for the Eradication of Poverty (1997–2006)*, Resolution A/RES/58/222 adopted by the General Assembly on the Report of the Second Committee (A/58/488), fifty-eighth session of the General Assembly, Agenda item 98 (published online 23 December 2003) <http://www.undemocracy.com/A-RES-58-222.pdf>.

Von Pischke, J.D. (1991), *Finance at the Frontier: debt capacity and the role of credit in the private economy* (Washington DC: The World Bank).

Von Stauffenberg, D., Jansson, T., Kenyon, N. and Barluenga-Badiola, M.-C. (2003), *Performance Indicators for Microfinance Institutions: technical guide* (3rd edition, Washington DC: MicroRate and Inter-American Development Bank) (published online) <http://www.iadb.org/sds/doc/52332techguid.eng.3v.pdf>.

Waterfield, C. and Ramsing, N. (1998), 'Management Information Systems for Microfinance Institutions: a handbook', *CGAP Technical Tool No. 1* (Washington, DC: CGAP/World Bank) (published online February 1998) <http://www.microfinancegateway.org/files/1631_044.pdf>.

Woller, G.M., Dunford, C. and Woodworth, W. (1999), 'Where to Microfinance?', *International Journal of Economic Development* 1:1, 29–64.

Woolcock, M. and Narayan, D. (2000), 'Social Capital: implications for development theory, research and policy', *The World Bank Research Observer* 15:2, 225–49.

Workgroup on Solidarity Socio-Economy (n.d.a), [website], <http://www.socioeco.org>.

Workgroup on Solidarity Socio-Economy (n.d.b), *Solidarity Finance* [website], <http://finsol.socioeco.org/en/>.

Yunus, M. and Jolis, A. (1998), *Banker to the Poor: the autobiography of Mohammad Yunus, founder of the Grameen Bank* (London: Aurum Press Ltd).

Chapter 4

Civil Society, Governmentality and the Contradictions of Governance-beyond-the-State: The Janus-face of Social Innovation

Erik Swyngedouw

Socially Innovative Governance: Towards a New Governmentality?

In recent years, a proliferating body of scholarship has attempted to theorize and substantiate empirically the emergence of new formal or informal institutional arrangements that engage in the act of governing outside, and beyond, the state (Rose and Miller 1992; Mitchell 2002; Jessop 1998; Pagden 1998; UNESCAP 2004; Whitehead 2003; Papadopoulos and Warin 2007). While much of this analysis of a changing, if not new, governmentality (or governmental rationality [Gordon 1991]) starts from the vantage point of how the state is reorganized in order to respond to changing socio-economic and cultural conditions and social demands for enlarged public participation, this chapter seeks to assess the consolidation of new forms of innovative governance capacity and the associated changes in governmentality (Foucault 1979) in the context of the rekindling of the governance–civil society articulation that is invariably associated with the rise of a neoliberal governmental rationality and the transformation of the technologies of government. In sum, the range of socially innovative forms of territorial development and governance are decidedly Janus-faced. While some can be emancipatory, inclusive and democratizing, others signal a more disturbing tendency towards the erosion of democratic accountability and the further consolidation of a fast-forwarding neoliberalization process. It is this double-edged character of socially innovative forms of governance and development that this chapter seeks to address.

In the context of this contribution, governance as an arrangement of governing beyond the state (though often with the explicit inclusion of parts of the state apparatus) refers to the proliferation of often socially innovative institutional or quasi-institutional arrangements of governance that are organized as associational networks of private, civil society (usually NGOs) and state actors. These forms of apparently horizontally organized ensembles are increasingly prevalent in rule making, rule setting and rule implementation at a variety of levels (Marcussen and Torfing 2007). Such arrangements of governance operate at various interrelated geographical scales, from the local/urban level (such as development corporations, *ad hoc* committees, stakeholder-based formal or informal associations dealing with social, economic, infrastructural, environmental or other matters) to the transnational scale (such as the European Union, the WTO, the IMF or the World Bank) (Swyngedouw 1997). These modes of governance have been depicted as a new form of governmentality, that is, 'the conduct of conduct' (Foucault 1982; Lemke 2002), in which a particular rationality of governing is combined with new technologies and instrumentalities of government.

The urban scale has been a pivotal terrain where these new arrangements of governance have materialized in the context of the emergence of innovative social movements on the one hand and transformations in the arrangements for conducting urban governance on the other (Le Galès 1995; Brenner and Theodore 2002; Jessop 2002a; Gonzalez and Healey 2005; Swyngedouw 2007). The main objective of this chapter is to address and problematize political citizenship rights and entitlements and to tease out the contradictory Janus-faced character that these newly emerging forms of governing the urban might have in terms of their democratic legitimacy. Indeed, the inclusion of civil society organizations (like NGOs) and other 'stakeholders' in systems of urban governance, combined with a greater political and economic role of 'local' political and economic arrangements, is customarily seen as potentially empowering and democratizing (Le Galès 2002), promoting innovative and creative adaptation through horizontal networking. While Governance-beyond-the-State may indeed embryonically contain elements that may permit greater openness, inclusion and empowerment of hitherto excluded or marginalized social groups, there are equally strong processes at work pointing in the direction of a greater autocratic governmentality (Swyngedouw 1996; 2000) and an impoverished practice of political citizenship. Such participatory networked forms of governance that centrally revolve around the mobilization of 'civil society' organizations within arrangements of governing, are actively encouraged and supported by agencies pursuing a neoliberal agenda (like the IMF or the World Bank) and 'designate the chosen terrain of operations for NGOs, social movements, and "insurgent" planners (see Sandercock 1998)' (Goonewardena and Rankin 2004: 188).

In the first part of the chapter, I outline the contours of Governance-beyond-the-State. Subsequently, I address the thorny issues of the state/civil society relationship in the context of a predominantly market-driven and neoliberal political–economic societal framework. The transformation of the state/civil society relationship will be situated within an analysis of consolidating neoliberal capitalism (Hardt and Negri 2000). In a third part, I tease out the contradictory ways in which new arrangements of governance have created innovative institutions and empowered new actors, while disempowering others. I argue that this shift from 'government' to 'governance' is associated with the consolidation of new technologies of government (Dean 1999) on the one hand, and with a transformation of the basic rules of deliberative democratic governing on the other. As such, this mode of governance entails both a transformation of the institutions and of the mechanisms of participation, negotiation and conflict intermediation (Coaffee and Healey 2003). Participation, then, is one of the key terrains over which battles about the form of governance and the character of regulation are currently being fought (Docherty et al. 2001; Raco 2000). I conclude by suggesting that socially innovative arrangements of Governance-beyond-the-State are fundamentally Janus-faced.

Governance-beyond-the-State: Networked Associations

It is now widely accepted that the system of governing within the European Union and its constituent parts is undergoing rapid change (European Commission 2001; Le Galès 2002). Although the degree of change and the depth of its impact are still contested, it is beyond doubt that the nineteenth-/twentieth-century political formations of articulating the state/civil society relationship through different forms of representative democracy, which vested power in hierarchically structured transcendental state-forms, is complemented by a proliferating number of new institutional forms of governing that exhibit rather different characteristics (Jessop 1995; Kooiman 1995; 2003). In other words, the Westphalian state-order that matured in the twentieth century in the form of the liberal–democratic state, organized at local, and often also at regional and national scales, has begun to change in important ways, resulting in new forms of governmentality, characterized by a new articulation between state-like forms (such as, for example, the EU, urban development corporations and the like), civil society organizations and private market actors (Brenner et al. 2003). While the traditional state-form in liberal democracies is organized through forms of political citizenship which legitimize state power by vesting it with the political voice of the citizenry, the new forms of governance exhibit a fundamentally different articulation between power and citizenship and, consequently, constitute a new form of governmentality. As Schmitter (2002, 52) defines it:

> Governance is a method/mechanism for dealing with a broad range of problems/ conflicts in which actors regularly arrive at mutually satisfactory and binding decisions by negotiating with each other and cooperating in the implementation of these decisions.

From this perspective, it is not surprising to find that such modes of 'Governance-beyond-the-State' are resolutely put forward as idealized socially innovative normative models (see Le Galès 1995; Schmitter 2000; 2002) that promise to fulfil the conditions of good and democratic government (European Commission 2001) 'in which the boundary between organizations and public and private sectors has become permeable' (Stoker 1998, 38). It implies a common purpose, joint action, a framework of shared values, continuous interaction and the wish to achieve collective benefits that cannot be gained by acting independently (Stoker 1998; Rakodi 2003). This model is related to a view of 'governmentality' that considers the mobilization of resources (ideological, economic, cultural) of actors operating outside the state system (hence the focus on 'civil society') as a vital part of democratic and effective government (Pierre 2000a; 2000b; Sørenson and Torfing 2003). These new forms of governing are based on horizontal interaction among presumably equal participants with no distinction between their public or private status, on regular and iterative exchanges among a set of independent but interdependent actors, and on improved access to the decision-making cycle for those who are customarily excluded from direct participation in decision-making and rule-setting procedures.

State-based arrangements are hierarchical and top-down (command-and-control) forms of setting rules and exercising power (but recognized as legitimate via socially agreed conventions of representation, delegation, accountability and control), and mobilizing governmental technologies marked by policing, biopolitical knowledge and bureaucratic rule. Governance-beyond-the-State systems, by contrast, are presumably horizontal, networked, and based on interactive relations between actors who share a high degree of trust, despite internal conflict and oppositional agendas, within inclusive participatory institutional or organizational associations. The mobilized technologies of governance revolve around reflexive risk-calculation (self-assessment), accountancy rules and accountancy-based disciplining, quantification and bench-marking of performance (Dean 1999; Burchell 1993), and entail the explosion of audit activity as a central disciplining and controlling technology (Power 1997).

The participants in such forms of governance partake (or are allowed to partake) in these networked relational forms of decision making on the basis of the 'stakes' they hold with respect to the issues these forms of governance attempt to address. Of course, such an idealized–normative model of horizontal, non-exclusive and participatory stakeholder-based governance is symptomatically oblivious to the contradictory tensions (the state/market/civil society conundrum) in which any form of governing is inevitably embedded. Before considering these contradictions, we need to turn our attention first to how these innovations in the arrangements of the 'conduct of conduct' parallel changing choreographies of civil society/state interaction, and, second, to the emergence of these socially innovative forms of governance in the context of broader processes of political–economic regime change.

State, Civil Society and Governmentality

There is considerable debate about the status, content and even definition of 'civil society', both analytically and empirically. This confusion arises partly from the meandering history of the concept. While the early Enlightenment view of 'civil society' posited 'civil' society versus 'natural' society, Hegel and Marx considered civil society as a set of economic/material relations that are positioned versus the state. Of course, this change in perspective was in itself related to the changing position of the state (from 'sovereign' to biopolitical; that is, from a state focused on the integrity of its territorial control to one operating allegedly in the 'interest of all for the benefit of all'). Liberal thinkers, such as de Tocqueville, in turn associated 'civil society' with voluntary organizations and associations. By contrast, with Antonio Gramsci, writing at a time of the formation of the Keynesian welfarist state, civil society became viewed as one of the three components (the others being the state and the market) that define the content and structure of, and contradictions in, society (Gramsci 1971).

In what follows, I take a Gramscian perspective, which – I believe – argues theoretically most systematically and convincingly for the continuing relevance of a notion of civil society (see also Poulantzas 1973; 1980; Jessop 1982; 2002b). For these authors, civil society both emerges, and cannot be theorized independently, from the way the (modern) liberal–democratic state operates (that is, its governmentality) on the one hand, and from the contradictory dynamics of societies in which capitalist social relations of production are dominant if not hegemonic on the other. Indeed, with the rise of the liberal state in the eighteenth century, civil society became increasingly associated with the object of state governing, as well as being the foundation from which the state's legitimacy was claimed. In addition, as the state became increasingly biopolitical, that is, concerned with, and intervening in, the 'life qualities' of its citizens (health, education, socio-economic well-being, among others), civil society emerged as both an arena for state intervention and a collection of actors engaging with, acting upon, and relating to the state (Lemke 2001). At the same time, the liberal state maintained the 'economic' sphere as fundamentally a 'private' one, operating outside the collective sphere of the state, but shaping the material conditions of civil life in a decisive manner. The social order, consequently, became increasingly constructed as the articulation between state, civil society and market. While for Hegel and Marx, albeit in very different ways, the ideal of society resided in transcending the separation between the 'political state' and 'civil society', the operation of the economy under the hidden hand of the market in liberal–capitalist societies rendered this desired unity of state and civil society impossible. In fact, a fuzzy terrain between state and market was produced, a terrain that was neither state nor private, yet expressed a diverse set of social activities.

It is not surprising that Gramsci struggled with the notion of civil society at a time of tumultuous political and economic change. The rise of a range of socially innovative experiments (collectives, cooperatives, anarchist alternatives, fascist militia and the like) in the 1920s and early 1930s, and the gradual emergence and consolidation of a Keynesian Welfare State form of liberalism, intensified qualitatively the interweaving of state, civil society and the market, but under the bureaucratic, hierarchical auspices of the state. According to the Gramscian perspective, the sum total of society is composed of three interrelated 'moments' or 'instances': a) the dynamics of dominant capitalist social relations (in the 1930s defined as the rise of 'Fordism'); b) the role and position of the state therein; and c) the heterogeneous, kaleidoscopic, dynamic, often fragmented, internally contradictory collective of actors who operate outside both the state and the capital valorization process, but who take specific positions with respect to both the state and the capital valorization process (that is, civil society). These include organizations like the church, unions, employers' federations, Masonic lodges, interest groups, the media, fascist militia and the like. Civil society comprises, consequently, a heterogeneous, conflicting and not necessarily 'civil' or 'democratizing' array of social actors (Edwards 2002). A variety of intellectuals play decisive roles in shaping and organizing the views, perspectives, and aspirations (that is, ideology) held within various parts of civil society (Gramsci 1971). Each of these civil society actors operates within the interstices and contradictory dynamics of capitalist society, while articulating activities in dialogue, confrontation and negotiation with, or parallel to, the state. In other words, civil society can neither be theorized independently from the content, form and structure of the state nor from the conjunctural characteristics of capitalism.

The state is inevitably and necessarily an ambiguous and internally contradictory institution under capitalism. On the one hand, the state is ideologically legitimized as the political expression of civil society. On the other hand, it has to sanction the private ownership of the means of production and guarantee the appropriation of surplus value in ways that support accumulation. Both conditions are extraordinarily fragile in their own right and often mutually contradictory. The first condition of the position of the state as the political expression of civil society, and realized within liberal democracy, reflects the embodiment of the sovereign will of the people. When the forces operating within civil society find their political expression within the coalition of forces that occupy the state apparatus, the relations between state and civil society are less oppositional. It is this condition that Gramsci associates with hegemony, a situation whereby the alliance of political forces (a political bloc) in power (that is, in control of the state) pursues a political programme broadly in line with, and reflecting, the majority aspirations, views and perspectives within civil society. The notion of governmentality is closely associated with this state/civil society relationship in the sense that the possibility of hegemony and, therefore, of governability (Donzelot 1991; Pagden 1998), is predicated upon a close correspondence between the dominant forces at work in civil society and those within the state apparatus. For example, in the 1920s and 1930s, unions (not merely defined as defenders of workers' economic interests – in many European countries unions were also powerful actors within the domains of culture, education, health services and the like) and other civil society organizations increasingly became directly entangled with the state apparatus, and a system of governmentality was then constructed – together with a series of '*dispositifs*' of governing usually referred to as corporatism – which welded together state and sectors of civil society in a fragile, but nevertheless more or less coherent, framework of government (Offe 1984). This ensemble would turn into one of the defining moments in the construction of hegemony (the correspondence between state and civil society) under Fordism, with its particular rationalities and technologies of power. Consequently, for Gramsci as well as for many other theorists of social change, creative innovative social and political conditions (of which, of course, social revolution is the most radical example) arise in a context of a widening gap between state and civil society, and at moments when the manifold and variegated demands, aspirations and activities operating within society (but outside the state) no longer find sufficient legitimate representation within the state. In other words, social change and social innovation entail changes in the state/civil society/market articulation. The legitimacy of the state diminishes of course to the extent that the act of governance (that is, the regulation of social and biological life) is organized through non-state actors; that is, it is provided by, and through, civil society. The destatization (Jessop 2002b) of a series of former state domains and their transfer to civil society organizations redefines the state/civil society relationship and structures a widening gap between state and civil society actors.

This contradictory process is paralleled by a second condition that pertains to the position of the state. To the extent that capitalist social relations are the dominant forms for the material organization of social life, the state has to guarantee these conditions primarily to secure and sanction private property rights and their associated institutions; and secondly, to regulate the appropriation of surplus in ways that support further accumulation. Notwithstanding the highly diverse possibilities for achieving this, the state, in pursuing a 'developmentalist' agenda that prioritizes market-organized economic growth above all other possible political objectives, still finds itself potentially in serious conflict with more or less significant parts of civil society. Those refusing to accept these conditions have to be excluded from the capitalist state (compare the exclusion of communists from the state apparatus during most of the post-war period in the West). But even amongst those who accept or defend capitalist social relations, there is considerable competition and tension, particularly concerning the conditions of accumulation and the mechanisms of appropriation and distribution of surplus value. It is also with respect to these tensions that the state finds itself in an inherently contradictory situation.

To sum up, while the state is a pivotal institution in the maintenance of social cohesion and legitimacy, it is continuously confronted with the requirements and limitations imposed by both the particular forms of capitalist social relations prevalent at any given time and the actions of social actors operating outside the state apparatus. Although liberal democracy claims that the place of power is structurally empty, only occupied temporarily by those representing 'the people' (see Lefort 1989), the maintenance of capitalist social relations demands the securing of a relative fixity of uneven power positions.

It must be evident from the above that 'civil society' is neither good nor bad (like the state!). It is internally lacking in homogeneity and, although collective, cannot be identified with the 'common' good (civil society organizations customarily pursue partial, positioned objectives). At the same time, 'civil society' is the actual concretely lived body of society; it is the sum total of actors, processes and resources mobilized outside the state and outside the dominant capital valorization process.

Consideration of the position of 'civil society' in relation to the state and the market is a fundamental vector for analysing particular historical–geographical conjunctures. It is exactly here that 'innovation' in a social and political sense takes place. It is of course also here that the actual forms of both state and market are contested, and the dynamics for change emerge (in a direction that can be either 'progressive' or 'regressive'). This analysis permits us to examine the growing attention paid, both theoretically and in political praxis, to 'civil society'. That is what I turn to next.

The Rise of 'Civil Society' and the Changing Regime of 'Governance'

As in Gramsci's time, academic and political attention has recently turned to 'civil society' at a point of profound restructuring of the political economy of capitalism and the concomitant crisis of the state. The rising importance of 'civil society' has, therefore, to be understood in the context of the changing relationship between state and economy on the one hand, and state and civil society on the other. A close association of state, market and civil society organizations by means of primarily nationally organized institutional configurations and regulatory procedures characterized the post-war period. These relationships have gradually broken down and started to develop in increasingly contradictory ways. The main dynamics of this rearticulation of state–civil society–market relate to three tendencies. Firstly, a growing discontent with respect to the particular institutional configurations of 'Fordist' capitalism has emerged among parts of civil society from the 1960s onwards. Civil rights movements, environmental movements, 'third world' organizations, 'alternative' post-materialist socio-cultural movements, feminist organizations and urban social movements, among others, began to assert their presence, largely operating outside (and often in opposition to) the dominant corporatist state state–economy alliance (Cruikshank 1999). In other words, a series of counter-hegemonic social movements arose from within civil society. These movements and actors operated strategically in the cracks and voids left by the then dominant political–economic regime and would eventually give rise to a gradually more active, activist and engaging civil society and to a growing and innovative 'social economy' or 'third sector' ensemble. The One of The People is increasingly replaced by the One of the Multitude (Virno 2004).

Secondly, there has been a great expansion of social, technical, spatial and cultural divisions within capitalism, which has generated both a much greater local differentiation (sectorally, organizationally, institutionally and so on) of forms of capitalist production and a greatly expanded transnational organization of capitalist markets and actors. In other words, the tenuous relationship between national territories and forms of accumulation or growth regimes began to give way to greater local differentiation and global integration. Processes of rapid deterritorialization accompanied radical forms of reterritorialization. This 'glocalization' of the economy redefined the relationship between state and economy in important ways (see Swyngedouw 1997; Brenner 2004).

Thirdly, the state itself is restructured. This, for our purposes, is the most important process for understanding the current position of 'civil society'. The 'crisis' of the state combines three interrelated dynamics:

- The rise of extra-parliamentary social movements that operate largely outside the state but to whose demands the state has to respond in one way or the other.
- The 'glocalization' of the economy which redefines the field of operation for the state and changes the 'constraints' under which the state operates in terms of maintaining or adjusting to the conditions that permit continuous accumulation.
- The internal crisis of the state, which is manifested in the twin pressure of mounting fiscal problems and increasing bureaucratization. The former, which is directly associated with the contradiction between changing economic parameters (which limit the fiscal room to manoeuvre) in a context of rising and unfulfilled demands from 'civil society', leads to a slimming down of the state's involvement with civil society through income redistribution and similar practices. The latter leads to deregulation and increasing removal of 'red tape', aimed at reducing state involvement and facilitating social action outside the state (Osborne and Gaebler 1993) and to a redrawing of the boundaries of the biopolitical state.

These forces combine in a reorganization of the state, while simultaneously redefining the contours of 'governmentality'. This reorganization takes three basic forms (Swyngedouw 1997; 2004):

- The externalization of state functions through privatization and deregulation (and decentralization): both mechanisms inevitably imply that non-state configurations become increasingly involved in regulating, governing and organizing a series of social, economic and cultural activities.
- The up-scaling of governance: the national state increasingly delegates regulatory and other tasks to other and higher scales or levels of governance (such as the EU, IMF, WTO and the like).
- The down-scaling of governance: greater local differentiation combined with a desire to incorporate new social actors in the arena of governing. This includes processes of vertical decentralization toward local governments. It can also take the form of de-responsibilization: transfer of functions to local communities without transfer of adequate resources/power.

These three processes of state 'glocalization' simultaneously reorganize the arrangements of governance as new institutional forms of Governance-beyond-the-State; these are set up, and become, part of the system of governing, of organizing the 'conduct of conduct'. Moreover, this restructuring is embedded in a consolidating neoliberal and consensual post-political ideological polity (see Swyngedouw 2007). The latter combines a desire to construct the market politically as the preferred social institution of resource mobilization and allocation, a critique of the 'excess' of state associated with Keynesian welfarism, a social engineering of the social in the direction of greater individualized responsibility, and the consolidation of new forms of networked and 'participatory' governance. Of course, the scalar reorganization of the state and the associated emergence of a neoliberal Governance-beyond-the-State redefine the state/civil society relationship in fundamental ways. The new articulations between state, market and civil society generate new forms of governance which combine the three 'moments' of society in new and often innovative ways. In particular, to the extent that both externalization and 'glocalization' of governing take place, other (often new) civil society organizations as well as private actors (stakeholders) become involved in the act of 'governing', either as a replacement of the state or in association with the state and/or the market.

However, the contested terrain of civil society initiatives and the content of 'governance' are subject to all manner of internal conflicts and tensions. The choreography of actual transformations in governance systems opens a vast arena of mutually interdependent mechanisms that significantly increase not only the complexity of the processes at work, but also, and perhaps more importantly, bring out the possible perverse effects or, at least, the contradictory character of many of these shifts. I turn to this next.

The Politics of Scale and Shifting Geometries of Power in Systems of Governance-beyond-the-State

As argued above, an important rescaling of the apparatuses of governance has taken place over the past few decades, combined with a growing externalization of public activities and functions. The 'traditional' local, regional and national state scales have increasingly positioned themselves offside as part of a strategy to recentre the regulatory force of the market as the main organizer of social relations (Jessop 2002b; Peck and Tickell 2002). Yet, this institutionalization of the 'market' as the principle organizing mechanism is paralleled by the emergence of new forms of 'governance' or of 'management'; as public–private networks composed of associated cultural, political and economic elites, these constitute important new domains of 'governing' at local, national and transnational scales (Swyngedouw et al. 2002). Even 'social economy' initiatives, together with NGOs of a variety of ideological stripes and colours, and other civil society organizations, partake in these fuzzy networks of governance to a greater or lesser extent (Kaika 2003; Kaika and Page 2003). The power relationships between citizens and governance shift, while, at the same time, the mechanisms of inclusion in and/or exclusion from these new forms of governance alter the choreography of power within civil society. There is a tendency towards loss of democratic control, while there is a corresponding growth in the power and influence of social and political–economic elites.

While such socially innovative networked forms of governance proliferate at a local level, a similar expansion of such forms of governance takes place at a supranational level. Although innovative in its own right, this process of 'glocalization' of governance in systems-beyond-the-state raises important questions with respect to the key practices through which 'democracy' and political citizenship rights and entitlements are organized. In lieu of the democratic representation that characterizes liberal democratic state forms, the new networked relations and formal or informal institutional ensembles are organized around interest groups of stakeholders. The status, inclusion or exclusion, legitimacy, system of representation, and internal or external accountability of such groups, or individuals, often takes place in non-transparent, *ad hoc* and context-dependent ways. The democratic lacunae of pluralist liberal democracy are well known, the procedures of democratic governing are formally codified, transparent and easily legible, but the *modus operandi* of networked associations are less clear. Moreover, the internal power choreography of systems of Governance-beyond-the-state is customarily led by coalitions of economic, socio-cultural or political elites. Therefore, the rescaling of policy transforms existing power geometries, resulting in a new constellation of governance characterized by a proliferating maze of opaque networks, fuzzy institutional arrangements, ill-defined responsibilities, and ambiguous political objectives and priorities. The 'glocalization' of governance is, therefore, often paralleled by a diminished sense of public involvement (unless via the intermediary of NGOs that have become stand-ins for segments of 'the people') and democratic content, by political exclusion and, consequently, by an uneven incorporation of sections of 'civil society' within these constellations of governance (for details, see Swyngedouw 2005). The latter do not imply a diminished position of the national state. On the contrary, recent research has shown conclusively that these new forms of glocal governance operate in close concert with both the private (market) sector and the state (Moulaert et al. 2002). In fact, it is the state that plays a pivotal and often autocratic role in transferring competencies (and consequently for instantiating the resulting changing power geometries) and arranging these new networked forms of governance. The democratic fallacies of the pluralist 'democratic' state are compounded by the expansion of the realm of 'governing' through the proliferation of such asymmetric Governance-beyond-the-State arrangements. To the extent that '*participation*' is invariably mediated by '*power*' (whether political, economic, gender or cultural) among participating 'holders', between levels of governance/government and between governing institutions, civil society and encroaching market power, the analysis and understanding of these shifting power relations are a central concern, particularly in light of the link between participation, social innovation and development (see Getimis and Kafkalas 2002).

The Janus-face of Governance-beyond-the-State: The Contradictions of Social Innovation in Governance

This chapter outlined the thesis of the transition in socio-economic regulation from statist command and control systems to horizontally networked forms of participatory governance. However, this thesis has to be qualified in a number of ways. First of all, the national or local state and its forms of political/institutional organization remain important. In fact, the state takes centre stage in the formation of the new institutional and regulatory configurations associated with governance. This configuration and associated techno-managerial '*dispositifs*' are directly related to the conditions and requirements of neoliberal governmentality in the context of a greater role of both private economic agents, as well as the more vocal civil-society-based groups. The result is a complex hybrid form of government/ governance (Bellamy and Warleigh 2001) in which the state externalizes some of its earlier authority and domains of intervention to non- or quasi-governmental associations.

Second, the characterization of socially innovative models of governance as non-hierarchical, networked and (selectively) inclusive forms of governmentality cannot be sustained uncritically. While governance promises and, on occasion, delivers a new relationship between the act of governing and society and, thus, redefines and reorganizes the traditional tension between the realization of the Rousseauan ideal in immanent forms of governing on the one hand, and the imposition of a transcendental Hobbesian leviathan on the other, there are also significant counter-tendencies. The up-scaling, down-scaling and externalization of functions traditionally associated with the scale of the national state have resulted in the formation of institutions and practices of governance that all express the above contradictions. This is clearly notable in the context of the formation (and implementation) of a wide array of socially innovative urban and local development initiatives and experiments on the one hand, and in the construction of the necessary institutional and regulatory infrastructure that accompanies such processes on the other. Needless to say, this ambiguous shift from government to a hybrid form of government/governance, combined with the emergence of a new hierarchically nested and articulated 'gestalt of scale', constitutes an important and far-reaching socio-political innovation.

Third, the processes of constructing these new choreographies of governance are associated with the rise to prominence of new social actors, the consolidation of the presence of others, the exclusion or diminished power position of groups that were present in earlier forms of government and the continuing exclusion of other social actors who have never been included. The new 'gestalt of scale' of governance, dubbed by some as post-political and post-democratic arrangements (see Swyngedouw 2007), has undoubtedly given a greater voice and power to some organizations (of a particular kind, for instance those who agree to play according to the rules set by the leading elite networks). It has consolidated and enhanced the power of groups associated with the drive towards marketization, and diminished the participatory status of groups associated with social–democratic or anti-privatization strategies.

Finally, and perhaps most important, Governance-beyond-the-State is embedded within autocratic modes of governing that mobilize technologies of performance and of agency as a means of disciplining forms of operation within an overall programme of responsibilization, individuation, calculation and pluralist fragmentation. The socially innovative figures of horizontally organized stakeholder arrangements of governance that appear to empower civil society in the face of an apparently overcrowded and 'excessive' state may, in the end, prove to be the Trojan Horse that diffuses and consolidates the 'market' as the principal institutional form.

In sum, these forms of governance, as Beck (1999, 41) argues, are full of 'unauthorized actors' and non-codified practices. While such absence of codification potentially permits and elicits socially innovative forms of organization and of governing, it also opens up a vast terrain of contestation and potential conflict that revolves around the exercise of (or the capacity to exercise) entitlements and institutional power. The democratic fallacies of the pluralist 'democratic' state are in fact compounded by the expansion of the realm of 'governing' through the proliferation of such asymmetric Governance-beyond-the-state arrangements.

References

BAVO (ed.) (2007), *Urban Politics Now: re-imagining democracy in the neoliberal city* (Rotterdam: NAI Publishers).

Beck, U. (1999), *World Risk Society* (Cambridge: Polity Press).

Bellamy, R. and Warleigh, A. (2001), *Citizenship and Governance in the European Union* (London: Continuum Publishers).

Brenner, N. (2004), *New State Spaces* (Oxford: Oxford University Press).

Brenner, N., Jessop, B., Jones, M. and MacLeod, G. (2003), 'Introduction: state space in question', in N. Brenner, B. Jessop, M. Jones and G. Macleod (eds), *State/Space: a reader* (Oxford: Blackwell).

Brenner, N. and Theodore, N. (2002), 'Cities and the Geographies of "Actually Existing Neo-Liberalism"', in Brenner, N. and Theodore, N. (eds), *Spaces of Neoliberalism* (Oxford: Blackwell).

Burchell, G. (1993), 'Liberal Government and Techniques of the Self', *Economy and Society* 22:3, 267–82.

Burchell, G., Gordon, C. and Miller, P. (eds) (1991), *The Foucault Effect: studies in governmentality* (Chicago, IL: The University of Chicago Press).

Coaffee, J. and Healey, P. (2003), '"My Voice: My Place": tracking transformations in urban governance', *Urban Studies* 40:10, 1979–99.

Cox, K. (ed.) (1997), *Spaces of Globalization: reasserting the power of the local* (New York: Guilford).

Cruikshank, B. (1999), *The Will to Empower: democratic citizens and other subjects* (Ithaca, NY: Cornell University Press).

Dean, M. (1999), *Governmentality: power and rule in modern society* (London: Sage).

Docherty, I., Goodlad, R. and Paddison, R. (2001), 'Civic Culture, Community and Citizen Participation in Contrasting Neighbourhoods', *Urban Studies* 38:12, 2225–50.

Donzelot, J. (1991), 'The Mobilization of Society', in Burchell et al (eds).

Dreyfus, H. and Rabinow P. (eds) (1982), *Michel Foucault: beyond structuralism and hermeneutics* (Brighton: Harvester).

Edwards, M. (2002), 'Herding Cats? Civil Society and Global Governance', *New Economy* 9:2, 71–6.

European Commission (2001), European Governance: a White Paper. COM(2001) 428 final (Brussels: Commission of the European Communities).

Foucault, M. (1979), 'On Governmentality', *Ideology and Consciousness* 6, 5–21.

Foucault, M. (1982), 'The Subject and Power', in Dreyfus and Rabinow (eds).

Getimis, P. and Kafkalas, G. (2002), 'Comparative Analysis of Policy-Making and Empirical Evidence on the Pursuit of Innovation in Sustainability', in Getimis, P., Heinelt, H., Kafkalas, G., Smith, R. and Swyngedouw, E. (eds) (2002), *Participatory Governance in Multi-Level Context: concepts and experience* (Opladen: Leske und Budrich).

Gonzalez, S. and Healey, P. (2005) 'A Sociological Institutionalist Approach to the Study of Innovation in Governance Capacity', *Urban Studies* 42:11, 2055–69.

Goonewardena, K. and Rankin, K.N. (2004), 'The Desire Called Civil Society: a contribution to the critique of a bourgeois category', *Planning Theory* 3:2, 117–49.

Gordon, C. (1991), 'Governmental Rationality', in Burchell et al. (eds).

Gramsci, A. (1971), *Selections from the Prison Notebooks* (London: Lawrence and Wishart).

Grote, J.R. and Gbikpi, B. (eds) (2002), *Participatory Governance: societal and political implications* (Opladen: Leske und Budrich).

Hardt, M. and Negri, A. (2000), *Empire* (Cambridge, MA: Harvard University Press).

Jessop, B. (1982), *The Capitalist State* (Oxford: Blackwell).

Jessop, B. (1995), 'The Regulation Approach: Governance and Post-Fordism: alternative perspectives on economic and political change', *Economy and Society* 24, 307–33.

Jessop, B. (1998), 'The Rise of Governance and the Risks of Failure: the case of economic development', *International Social Science Journal* 50:155, 29–46.

Jessop, B. (2002a), 'Liberalism, Neoliberalism and Urban Governance: a state-theoretical perspective', *Antipode* 34:2, 452–72.

Jessop, B. (2002b), *The Future of the Capitalist State* (Oxford: Blackwell).

Kaika, M. (2003), 'The Water Framework Directive: a new directive for a changing social, political and economic European framework', *European Planning Studies* 11:3, 299–316.

Kaika, M. and Page, B. (2003), 'The EU Water Framework Directive Part 1: European policy-making and the changing topography of lobbying', *European Environment* 13:6, 314–27.

Kooiman, J. (1995), *Modern Governance: new government-society interactions* (London: Sage).

Kooiman, J. (2003), *Governing as Governance* (London: Sage).

Lefort, C. (1989), *Democracy and Political Theory* (Minneapolis, MN: Minnesota University Press).

Le Galès, P. (1995), 'Du Gouvernement Local à la Gouvernance Urbaine', *Revue Française de Science Politique* 45, 57–95.

Le Galès, P. (2002), *European Cities: social conflicts and governance* (Oxford: Oxford University Press).

Lemke, T. (2001), '"The Birth of Bio-politics" – Michel Foucault's Lecture at the Collège de France on Neo-Liberal Governmentality', *Economy and Society* 30:2, 190–207.

Lemke, T. (2002), 'Foucault, Governmentality, and Critique', *Rethinking Marxism* 14:3, 49–64.

Marcussen, M. and Torfing, J. (2007), *Democratic Network Governance in Europe* (Basingstoke: Palgrave).

Mitchell, K. (2002), 'Transnationalism, Neo-Liberalism, and the Rise of the Shadow State', *Economy and Society* 30:2, 165–89.

Moulaert, F., Rodriguez, A. and Swyngedouw, E. (eds) (2002), *The Globalized City: economic restructuring and social polarization in European cities* (Oxford: Oxford University Press).

Offe, K. (1984), *Contradictions of the Welfare State* (London: Hutchinson).

Osborne, D. and Gaebler, T. (1993), *Reinventing Government: how the entrepreneurial spirit is transforming the public sector* (New York: Penguin).

Pagden, A. (1998), 'The Genesis of "Governance" and Enlightenment Conceptions of the Cosmopolitan World Order', *International Social Science Journal* 50:155, 7–15.

Papadopoulos, Y. and Warin, P. (2007), 'Are Innovative, Participatory and Deliberative Procedures in Policy Making Democratic and Effective?', *European Journal of Political Research* 46, 445–72.

Peck, J. and Tickell, A. (2002), 'Neoliberalizing Space', *Antipode* 34:3, 380–404.

Pierre, J. (ed.) (1998), *Partnerships in Urban Governance: European and American experience* (Basingstoke: Macmillan).

Pierre, J. (2000a), *Debating Governance: authority, steering and democracy* (Oxford: Oxford University Press).

Pierre, J. (2000b), *Governance, Politics and the State* (Basingstoke: Macmillan).

Poulantzas, N. (1973), *Political Power and Social Classes* (London: New Left Books).

Poulantzas, N. (1980), 'Research Note on the State and Society', *International Social Science Journal* 32:4, 600–608.

Power, M. (1997), *The Audit Society: rituals of verification* (Oxford: Oxford University Press).

Raco, M. (2000), 'Assessing Community Participation in Local Economic Development: lessons for new urban policy', *Political Geography* 19:5, 573–99.

Rakodi, C. (2003), 'Politics and Performance: the implications of emerging governance arrangements for urban management approaches and information systems', *Habitat International* 27:4, 523–47.

Rose, N. and Miller, P. (1992), 'Political Power beyond the State: problematics of government', *British Journal of Sociology* 43, 173–205.

Sandercock, L. (1998), *Towards Cosmopolis* (Chichester: J. Wiles and Sons).

Schmitter, P. (2000), 'Governance', Paper presented at the Conference of Democratic and Participatory Governance: From Citizens to 'Holders', European University Institute, Florence, Italy, 14 September.

Schmitter, P. (2002), 'Participation in Governance Arrangements: is there any reason to expect it will achieve "sustainable and innovative policies in a multi-level context"?', in Grote and Gbikpi (eds).

Sørenson, E. and Torfing, J. (2003), *Theories of Democratic Network Governance* (Basingstoke: Palgrave Macmillan).

Stoker, G. (1998), 'Public-Private Partnerships in Urban Governance', in Pierre (ed.).

Swyngedouw, E. (1996), 'Reconstructing Citizenship, the Re-scaling of the State and the New Authoritarianism: closing the Belgian mines', *Urban Studies* 33:8, 1499–521.

Swyngedouw, E. (1997), 'Neither Global nor Local: "glocalization" and the politics of scale', in Cox (ed.).

Swyngedouw, E. (2000), 'Authoritarian Governance, Power and the Politics of Rescaling', *Environment and Planning D: Society and Space* 18, 63–76.

Swyngedouw, E. (2004) 'Globalisation or "Glocalisation"? Networks, territories and rescaling', *Cambridge Review of International Affairs* 17:1, 25–48.

Swyngedouw, E. (2005), 'Governance Innovation and the Citizen: the Janus face of governance-beyond-the-state', *Urban Studies* 42:11, 1991–2006.

Swyngedouw, E. (2007), 'The Post-Political City', in BAVO (ed.).

Swyngedouw, E., Moulaert, F. and Rodriguez, A. (2002), 'Neoliberal Urbanization in Europe: large-scale urban development projects and the new urban policy', *Antipode* 34:3, 542–77.

UNESCAP (2004), 'What is Good Governance?', United Nations Economic and Social Commission for Asia and the Pacific, <http://www.unescap.org/huset/gg/governance.htm>, accessed 9 March 2004.

Virno, P. (2004), *A Grammar of the Multitude* (Cambridge, MA: MIT Press).

Whitehead, M. (2003), '"In the Shadow of Hierarchy": meta-governance, policy reform and urban regeneration in the West Midlands', *Area* 35:1, 6–14.

PART II
Cities and Socially Innovative Neighbourhoods

Chapter 5

Social Innovation for Neighbourhood Revitalization: A Case of Empowered Participation and Integrative Dynamics in Spain

Arantxa Rodriguez

Introduction

Increasing inequality and social exclusion has become an alarmingly conspicuous feature of today's urban landscapes (Mingione 1996; Martens and Vervaeke 1997; Pacione 1997; Wilson 1997; Marcuse and van Kempen 2000; Moulaert, Morlicchio and Cavola 2007). In cities throughout the world, this trend follows the reorganization of global competitive conditions and national welfare regimes that, since the mid 1970s, have drastically reshaped labour market and income opportunities as well as access to housing and services. In the brave new economy of post-Fordist accumulation, social exclusion tensions have intensified for an expanding and ever more vulnerable part of society through the spread of casualized forms of employment, weakening social protection systems and increasing individualization of economic risk and uncertainty (Lipietz 1989; European Commission 1997; Jessop 1994).

The expansion of social exclusion has been accompanied by the increasing concentration of disadvantage in particular urban areas reinforcing existing patterns of spatial segregation and engendering new socio-spatial divisions (Pacione 1997; Atkinson and Kintrea 2001; Forrest and Kearns 1999; Glennerster et al. 1999). The persistence and extension of deprived neighbourhoods both in dynamic cities as well as less prosperous ones reflect the enduring spatiality of social exclusion, that is, the necessary materialization in space of structural, macro-scale processes shaped by localized factors that operate through labour, land and property markets and by local regulatory and governance structures (Moulaert et al. 1994; Moulaert et al. 2000; Moulaert et al. 2003; Allen et al. 2000). In disadvantaged neighbourhoods, the spatial concentration of poor households and individuals interacts with factors such as poor housing, neglected infrastructures and services, lack of job opportunities, high crime and insecurity levels, physical isolation, institutional oblivion, and so on, exacerbating processes of social, economic, cultural and political exclusion that breed the formation of 'excluded communities' (Geddes 1997).

Increasing recognition of widening inequalities and awareness of the spatiality of disadvantage have contributed to a renewed interest in the role of place-based factors and neighbourhood effects in shaping processes of social exclusion and inclusion (Madanipour et al. 1998; Murie and Musterd 2004) as well as shedding new light on the long-standing debate about how best to tackle the needs of disadvantaged areas. Over the last two decades, the neighbourhood has been rediscovered as a relevant analytical and policy scale and, in many cities, it has become the primary battleground against social exclusion (Kearns and Parkinson 2001; Parkinson 1998; Hull 2001; Morrison 2003). Indeed, the current focus on neighbourhoods incorporates innovative policy approaches that stress greater policy integration and coordination together with a rejuvenated emphasis on joined-up solutions, community participation, partnership and empowerment as a means of improving policy delivery and outcomes, promoting political inclusion and strengthening local democracy (Raco and Imrie 2003; Amin 2005). Deprived urban communities have, in this way, become key strategic sites in the changing urban policy arena.

This chapter explores these challenges in relation to recent changes in neighbourhood regeneration policies in Spain and the implications for developing integrative dynamics in deprived neighbourhoods. First, the chapter examines the re-emergence of neighbourhood as a relevant analytical and policy scale. Second, it discusses changing policy agendas in neighbourhood regeneration. Third, it presents the evolution of neighbourhood policies in Spain and the emergence of alternative, innovative schemes during the last decade. Finally, it discusses social innovation in neighbourhood regeneration in Trinitat Nova.

Neighbourhood Regeneration, Inclusion and Social Innovation

Public intervention in deprived neighbourhoods has been an integral part of urban policy in many European and North American cities throughout the twentieth century (OECD 1998). However, planned neighbourhood regeneration programmes were launched mostly in the post-war period in the context of intense industrialization and urban development. Since then, public intervention in distressed urban areas has evolved in response to changing socioeconomic and political dynamics at different territorial scales as well as to the changing understanding of social and spatial disadvantage. Notwithstanding significant variation in the specific content and timing across countries, three major waves of neighbourhood-based initiatives have been identified (Carmon 1997).

The first wave of deliberate public intervention in deprived neighbourhoods can be traced back to massive slum clearance initiatives implemented in the 1930s and 1940s in Britain and the United States. In the post-war years, slum clearance programmes were ostensibly portrayed as a means to provide decent and affordable housing for poor residents in rundown areas and tied to national housing legislation and the construction of large public housing projects. Strongly biased by a sectoral, top-down and unequivocally physical approach, the capacity of these programmes to improve the living (or housing) conditions of the poor was extraordinarily limited (Keating et al. 1996). Still, in Britain, demolition of slum areas came simultaneous with large-scale production of council housing, allowing local authorities to comply with the statutory obligation to relocate displaced households. In the United States, however, slum clearance aimed at revival of inner city areas adjacent to downtown, and encouraged private redevelopment for more productive economic uses, new infrastructures and sometimes also higher income housing, forcing low-income minority residents to relocate into similar deprived neighbourhoods (Keating and Smith 1996). Similar schemes were implemented in other advanced industrial economies, such as France, Canada and the Netherlands, in the name of modernization, reconstruction and reducing acute housing shortages.

By the late 1960s, slum clearance programmes came up against intense criticism and resistance because of their high social and economic costs as well as their limited capacity to actively improve welfare conditions for residents in rundown neighbourhoods. Public intervention in distressed neighbourhoods began to shift away from demolition and urban renewal towards an emphasis on improving housing and neighbourhood conditions through renovation initiatives; rehabilitation rather than demolition became the main approach to neighbourhood regeneration as the strong physical bias gave way gradually to a greater concern with poverty and the consequences of changing socio-spatial inequalities resulting from fast-paced economic growth and urbanization. The new emphasis on improving housing and physical environment in deprived neighbourhoods ran parallel to growing opposition to clearance and large-scale urban renewal and infrastructure developments while the search for more comprehensive, focalized and inclusive approaches emerged as a salient feature of second wave neighbourhood intervention policies in cities on both sides of the Atlantic (Couch et al. 2003).

By the end of the 1970s, economic crisis and restructuring processes paved the way for a new turn in neighbourhood regeneration programmes in which social and spatial disadvantage issues were largely neglected or subordinated to the imperatives of economic recovery, job creation and competitiveness. In the next two decades, the search for growth and competitive restructuring not only dominated the urban policy agenda but also led the transformation of local governance towards more entrepreneurial approaches to regeneration. Large-scale, flagship urban redevelopment projects were the most ubiquitous tool of entrepreneurial regeneration in European and North American cities (Swyngedouw et al. 2002; Moulaert et al. 2003). The logic of market efficiency, feasibility and competitiveness was firmly upheld as mainstreamed neighbourhood regeneration schemes were instrumentalized to serve the aims and means of the New Urban Policy agenda (Rodríguez and Vicario 2005).

Nevertheless, urban socio-spatial segregation, marginalization and disadvantage issues were not fully annihilated from the research and policy agenda of cities and social renewal schemes and neighbourhood-based integrated approaches coexisted – even if largely overlooked – with large-scale urban redevelopment projects in most Western economies (Parkinson 1998). By the 1990s, intensifying levels of inequality and a notorious lack of spillover effects from flagship project redevelopment onto poor neighbourhoods reinstated the need for specific, targeted programmes to combat social exclusion in disadvantaged neighbourhoods and urban areas (OECD 1998).

Mainstreaming Regeneration in Deprived Neighbourhoods

In this period, increasing recognition of intensifying social and spatial inequalities, as well as the failure of 'trickle down' benefits to reach deprived neighbourhoods, reintroduced problems of social exclusion and deprivation and the need for innovative actions to combat social exclusion in disadvantaged areas in the urban policy agenda. Despite critical variations in timing, content and character of these initiatives, most large European cities adopted a comprehensive, targeted, area-based approach to neighbourhood regeneration. The new turn – or rediscovery, in some cases – towards area-based strategies aimed to move beyond the largely physical, housing-based and property-led policies of previous neighbourhood regeneration rounds by addressing other critical dimensions of neighbourhood development, notably economic and social.

In France, integrated area-based initiatives were introduced in the early 1980s, in response to riots and social unrest in peripheral housing estates in Paris, Lyon and other major cities, through the neighbourhood social development policy, *Développement Social des Quartiers*, and later improved through the *Contrats de Ville*, a formalized agreement between national government and local authorities to tackle social exclusion at neighbourhood level (Anderson 1998). But in the late 1990s, neoliberalizing pressures largely neutralized these schemes. In Britain, New Labour drew upon a long tradition of area-based initiatives going back to the 1960s 'inner-city' policies to launch in 1998 *The New Deal for Communities*, a concerted, joined-up, place-based strategy for urban regeneration in the context of a wider national policy for combating social exclusion. Strategic elements and key innovations of this approach were subsequently mainstreamed into broader public policies and the *Neighbourhood Renewal Fund* (Hull 2001). In the Netherlands, integrated regeneration approaches developed as part of efforts to deal with spatial concentration of social problems in particular areas in the late 1980s but especially after the establishment of the *Strategic Neighbourhood Approach* within the 'Large Cities Policy' in the mid 1990s (Priemus 2005). Integrated area initiatives have also been implemented in Germany, Denmark, Belgium, Italy and, to a lesser extent, in other European Union countries to promote social inclusion and cohesion. Thus, the adoption of area-based approaches not only took rather different timing paths across countries and cities, but the way in which these initiatives evolved and their particular content and orientation was also very diverse (Van den Berg et al. 1998).

At the European scale, the area-based approach was also adopted in the 1990s as the main basis for community intervention in deprived urban areas and incorporated within the EU mainstream regional development policy through the *Urban Pilot Programme* and the *URBAN Community Initiative (1994–1999).*[1] Still, the need for a more focused and targeted approach to urban intervention was not openly laid out until 1997 in the Commission's Communication *Towards an Urban Agenda in the European Union* (COM (97)197), which led to the *European Union Framework for Action for Sustainable Urban Development* (COM/1998/605), the main policy tool for promoting social inclusion and regeneration in urban areas through integrated, area-based and partnership-led initiative bringing together various policy domains, government levels, institutional actors and social networks. These initiatives were very influential in promoting the adoption of innovative area-based approaches to regeneration in deprived neighbourhoods in many cities through their funding by EU Structural Funds, a top-down enticement to the mainstreaming of bottom-up, integrated and participatory neighbourhood regeneration.

In sum, as social exclusion and inclusion issues have been moving up the urban policy agenda in the 1990s and 2000s, the policy focus for improving deprived neighbourhoods has been largely localized through the mainstreaming of integrated area-based approaches. Yet, despite the growing influence of the neoliberal urban agendas, area-based approaches remain a relevant arena for initiatives leading to social inclusion, particularly in a context of changing urban governance and the greater emphasis on strong partnership and concerted action, the building of local capacity and the active engagement of disadvantaged groups and local residents (Healey 1998). Indeed, in the new regeneration orthodoxy, the disadvantaged neighbourhood thus emerges as a prime site of participatory governance.

Civic mobilization and engagement emerge, in this context, as a means of local empowerment as well as greater cooperation and coordination among public, private and community sectors, while local partnerships are projected as the strategic means for enhanced efficiency in fighting social exclusion through the mobilization of all relevant stakeholders (Healey 1998). Community involvement may take quite different forms according to the action and the possibilities of actors to influence outcomes. Community participation may be merely a means to rearticulate a regime of efficient policy delivery, self-regulation and transferred responsibility at the neighbourhood scale (Raco 2000; Raco and Imrie 2003; Peck

---

1 The *Urban Pilot Programme* was launched in 1990 for a three-year period to support innovative projects in urban regeneration and planning and later extended for a second phase (1997–1999). The *URBAN* initiative built upon the experiences of the Pilot projects and the *Quartiers en Crise* programme to promote regeneration of deprived urban areas through integrated, neighbourhood-based and community-led partnerships (Jacquier 1990).

and Tickell 2002; Swyngedouw 2005; Leitner et al. 2007) or, on the contrary, it can be an opportunity for the self-organization of socially excluded groups to gain control over the aims and priorities of intervention (Kearns and Paddison 2000; Healey 1998). In the first case, civic engagement acts largely as a legitimizing instrument for predetermined actions; in the latter, it constitutes an end in itself, the basis for genuine empowerment, inclusion and radicalized local democracy. The extent to which participatory governance adopts one form or the other depends crucially on the way society-state and institutional relations are negotiated and reconstructed at the local scale.

Regarding the disadvantaged neighbourhood, its validation as a prime site of participatory governance is hardly self-evident. Indeed, the capacity of deprived locations to be meaningful arenas of civic involvement is contingent upon specific local conditions and institutional structures and dynamics at particular times. The possibilities for effective, non-instrumental participation of residents are generally predicated upon the existence of a critically conscious, organized and active local community. But integration of different social groups and residents can also be actively encouraged through participatory institutional arrangements that create favourable conditions for citizens to participate in, and to influence, policies that have a direct impact on their lives. And it is here that institutional innovation plays a critical role in setting a stage that is either favourable towards instrumentalized and co-opted participation or, on the contrary, supportive of effective capacity building and empowered participation (Fung and Wright 2003; Rodriguez and Abramo 2008). This tension is at the heart of creative and innovative policies and approaches to neighbourhood revitalization developed during the last two decades in many cities.

Social Innovation in Neighbourhood Regeneration: Radicalizing Local Democracy

Innovative community development initiatives carried out in predominantly disadvantaged neighbourhoods and localities have been a critical – if inconspicuous – part of the urban policy landscape. Indeed, deprived urban areas have often been seedbeds of experimentation and innovation in community development putting forward various types of initiatives aimed explicitly at promoting inclusion in different spheres of society: economic, social, cultural, political and so on. And, in contrast to the technocratic, serialized and instrumentalist innovations of the New Urban Policy, alternative community-led initiatives exhibit a palimpsest of extraordinarily creative, socially innovative and counter hegemonic development practices (Moulaert et al. 1994; Moulaert et al. 2000; Moulaert et al. 2005; SINGOCOM 2005).

Innovative approaches to combat social exclusion at the neighbourhood scale gained high visibility in the early 1990s as a result of the EU Poverty III Programme (1989–1994). The Poverty III Programme, mainly a policy-oriented action programme, also contained a research programme aimed at developing an expanded view of social exclusion processes combining income inequality and material exclusion dimensions with social and cultural exclusion. In this research, exclusion involves both distributional as well as relational dynamics that are brought together through the notion of citizenship rights[2] (Atkinson 2000). This view of exclusion encouraged a radical shift away from sectoral and fragmented schemes towards more integrated multi-agent, multi-agenda and multi-level governance approaches to neighbourhood and city development. Moulaert et al. (1990) advanced the notion of Integrated Area Development (IAD) to capture this view and define an alternative strategy for social inclusion in deprived areas.

In contrast to the construction of civic involvement and partnerships from above, IAD is an essentially bottom-up, community-led and multidimensional approach centred on the satisfaction of basic needs, economic and social mobilization, and political dynamics that allow the establishment of enabling institutions (Moulaert et al. 2000). IAD grounds itself in radically innovative social practices that establish social inclusion of the most vulnerable groups as a top priority of regeneration initiatives. As such, it moves beyond the narrow economistic bias of many local development strategies placing social and political integration goals side by side with economic integration goals in neighbourhood development planning.

Participation of local residents, community organizations and grassroots movements plays a key role in IAD approaches and is primarily oriented towards facilitating the revealing of needs. Participation is thus linked to a collective decision-making process involving meaningful communication and confrontation of interests between all relevant political subjects including all those concerned and affected by the decisions (Healey 1997). While decidedly challenging and non-instrumental, this form of citizen participation is actively supported through innovative institutional configurations reflecting the commitment of IAD approaches to community empowerment and a participatory urban democracy. In this way, IAD goes beyond the simple integration of policy domains, government levels and institutional actors, and social networks by re-centring the organic demands and the political and social organizations of excluded groups.

In the IAD framework, social innovation involves empowerment, and it is innovation in social relations that is the most relevant form of innovation for developing integrative local development dynamics. In this way, IAD is less prone to cooptation and instrumentalization by exogenous top-down policy agendas, thus transforming community engagement and participatory dynamics on the stepping-stones of a radical approach to neighbourhood revitalization.

---

2 Citizenship rights are embedded in societal institutions that provide the means for civic, economic, social and interpersonal integration. Social exclusion results when the breakdown of one or more of the systems of integration (the legal system, the labour market, the welfare system, and the family and community system) means that some individuals or groups cannot access the primary social, occupational and welfare institutions that allow the realization of citizenship rights.

Social innovation as empowered participatory governance draws the boundaries of innovative local democratic practices that attempt to restructure democratic decision-making, differentiating them from participative dynamics framed by the objectives of state agencies to enhance effectiveness. Empowered participatory governance refers to institutional reforms and experiences that are participatory because they rely upon the commitment and capacities of citizens to make decisions through deliberative processes, and empowered because they attempt to link action to discussion influencing policy outcomes (Fung and Wright 2003). In the context of neighbourhood regeneration policies, empowered participatory governance provides a measure against which the emancipatory potential of different experiences or institutional policies can be evaluated.

Social Innovation in Neighbourhood Regeneration in Spain

The dynamics of urban development and policy making in Spanish cities bear substantial differences *vis-à-vis* those of their northern European counterparts[3] as a result, largely, of late industrialization processes and an autocratic regime that lasted till the mid-1970s. Economic and urban growth lagged behind most advanced industrial economies before taking off at phenomenal rates in the early 1960s through the 1970s. Accelerated urbanization took place under conditions of significant economic and social hardship and political dictatorship, a context prone to unscrupulous practice and shortfalls. The downturn of the economic cycle brought to the forefront the limits of the 'developmentalist' urban model as problems arising from industrial decline and restructuring blended rather uneasily with mounting deficits in urban services and infrastructures produced by two decades of intense, largely unplanned and highly speculative urbanization, a challenge to which the urban policy framework had considerable difficulty adjusting (Terán 1999).

Urban problems gained increased visibility during the transition to democratic rule and became a locus of social and political mobilization. In many cities, an explicitly *urban* politics emerged as the privileged terrain of democratic reconstruction centred on living conditions and everyday life demands. Urban social movements that had often acted as surrogates for banned political parties during the dictatorship took the lead reclaiming the 'right to the city', integrating social justice and equity considerations in local planning agendas (Leal 1989). Not surprisingly, concrete material demands, such as improved access to urban services and social infrastructures, went hand in hand with strong claims of institutional reform in municipal governments and securing greater citizen participation in urban policy.

---

3   See Moulaert, Morlicchio and Cavola (2007) for a comparative analysis of northern and southern European cities' dynamics.

The Urban Master Plans approved in the early 1980s reflected the new social, economic and political priorities in cities; the significance of citizens' participation, attention to basic services and needs, a focus on deprived neighbourhoods and urban inequalities and a structural, integrated and coordinated approach to the city were common themes in these municipal plans (Leal 1989). The deceleration of economic and urban growth also contributed to shifting the focus of planning away from developmentalist schemes and projects towards upgrading and balancing the excesses of rapid urbanization through small-scale, remedial and qualifying interventions in public spaces, a trademark of what came to be known as 'austerity planning' (Campos Venuti 1981).

The first wave of neighbourhood regeneration schemes in Spain can be traced back to this period, targeting problems of physical dereliction, social disadvantage and marginalization in vulnerable districts and distressed neighbourhoods, notably historical city centre districts and peripheral housing estates (Arias 2000; Alguacil 2006; Pol 1988). Different regeneration initiatives and strategies at various administrative scales (local, regional, central government) addressed specific forms of spatial disadvantage but, historically, they have followed two distinct sectoral approaches: urbanistic and social. Urbanistic spatial planning initiatives targeted the physical renewal of housing, buildings and public spaces; social initiatives aimed at improving welfare conditions and the quality of life of deprived groups. These schemes were – and still are – implemented primarily by local and regional administrations but in coordination with central government guidelines. And, while most programmes targeting neighbourhood disadvantage were developed in isolation from one another, more integrated schemes targeting social exclusion in particular areas were also developed by central government agencies in the late 1980s and 1990s (Arias 2000).

One of the most relevant neighbourhood regeneration initiatives of this period was the *Neighbourhood Renewal Operation of Madrid*, a major project of transformation of the southern periphery of the city, consisting of the demolition of sub-standard housing built in the 1950s and 1960s and replacing it with new units and collective infrastructures and services in a total of 30 neighbourhoods (840 ha). The programme was launched in 1979 in response to demands from powerful grassroots organizations, which then played a leading role in the process of design and implementation in close partnership with the technical and political representatives from the regional administration. It concluded in 1996, after completing the construction of almost 39,000 new units purchased at a cost of less than 10 per cent of a family's income, rehousing 150,000 residents in the same location – a public investment of close to 1,900 million Euros (Ciudades para un Futuro más Sostenible n.d.a).

Similar initiatives were implemented in other major cities including Barcelona, Valencia and Seville. But in most cities, neighbourhood regeneration schemes were developed through 'Special Area Plans', a planning tool established specifically for dealing with area improvement. These plans were widely used during the 1980s in small- and medium-sized cities. A well-known example is the Rehabilitation

Plans for historical districts that flourished during that decade (Pol 1988; Bellet n.d.). Rehabilitation Plans were, by and large, intensely biased towards physical renovation of public spaces and buildings with little or no attention paid to social, economic or political issues. Indeed, an important lesson drawn from this first generation of neighbourhood renewal schemes in Spain was that, while these schemes were extremely effective in improving physical conditions, housing, infrastructure and urban services, they failed to take into account the social and economic dimensions necessary to tackle problems of unemployment, local capacity building and so on.

Attempts to overcome the strong physical bias of functionalist land-use planning and transcend the logic of urban architecture, housing policies and remedial initiatives were a driving force of neighbourhood regeneration schemes from the late 1980s. A move towards more comprehensive and integrated approaches was encouraged in the early 1990s through the Integrated Area Rehabilitation Act (ARI), established by the central administration in 1993 and developed in coordination with local and regional authorities. This programme was an important step forward in integrating different policy domains as well as institutional scales.

Another important thrust towards integrative approaches to social exclusion and spatial disadvantage in Spain was provided by the implementation of EU-funded programmes such as the Quartiers en Crise, Urban Pilot Projects and the URBAN initiative, but also through the deployment of projects financed by the European Social Funds to combat poverty in areas of high long-term unemployment. These programmes were often linked to local employment initiatives developed by many municipalities to cope with the consequences of economic and urban restructuring and improve the conditions for economic and urban redevelopment. This second wave of urban regeneration schemes was established mostly at the municipal scale as a tool for improving local production conditions and employability with a particular focus on training and the development of quality jobs, and the social economy, as well as a concern with wider issues of income distribution, social integration and community empowerment (Moulaert et al. 1994). However, the increasing mobilization of local politics in support of economic development gradually subordinated social policies to economic and labour market policies, stripping local economic development strategies of their more radical components.

The mainstreaming of local economic development strategies paved the way for the adoption of more strategic approaches in the early 1990s. In the context of widespread challenges to statutory planning and regulatory procedures and the spread of an anti-planning ideology, the focus of urban regeneration became the urban project. The project could be undertaken at various scales, from small 'acupunctural' type of interventions, through intermediate scale projects, to the more ambitious 'project of projects' or city-project (Busquets 1992). Yet, in Spain, as in many other European and North American cities, large-scale emblematic operations became the preferred mode of a new model of urban intervention. Project-led regeneration strategies were most powerfully expressed in the cases of

Barcelona, Seville and Madrid, followed closely by Bilbao, Valencia, Malaga and Oviedo (Rodríguez et al. 2001).

However, large-scale monumental urbanism and entrepreneurialism were by no means the only urban regeneration strategies pursued since the 1990s. Innovative neighbourhood regeneration schemes also evolved from initiatives developed in the context of new urban planning and implementation procedures integrating citizens' participation and coordination. An important trend in the new wave of neighbourhood revitalization, has been the increasing role played by regional governments, notably in Andalucía, Madrid, Catalonia and the Basque Country (Ciudades para un Futuro más Sostenible n.d.b), through the establishment of specific mechanisms, norms and tools for the development of integrated action in disadvantaged neighbourhoods.

A key initiative in the development of more integrative approaches has been the *Community Development Plan* launched by the Catalonian government in the late 1990s with the objective of, first, territorializing anti-poverty measures established in the *Integrated Plan for Combating Poverty and Social Exclusion*; second, propping up the *Community and Civic Participation Plan*; and, third, promoting citizen participation and associational dynamics (Blanco 2005). By the end of 1997, ten municipalities had produced Community Development Plans involving a total of 31 neighbourhoods. Changes in the political scene led to the gradual demise of this programme. However, new schemes have been developed in the 2000s along the same guidelines, notably the *Special Neighbourhood and Urban Areas Programme* launched by the Catalonian government in 2004, a pioneering initiative to combat social exclusion in disadvantaged neighbourhoods through integrated urban, social and economic regeneration (Generalitat de Catalunya n.d.).[4]

Finally, in recent years, an important wave of neighbourhood regeneration initiatives has sprung up from grassroots and civil society organizations struggling against intensifying social exclusion and disadvantage. They include an extremely rich spectrum of initiatives ranging from Community Plans at neighbourhood scale in Barcelona, Valencia, Pamplona, Madrid and others, to Participatory Intervention Nuclei (NIP), Strategic Plans, Participatory Budgeting in Córdoba, San Juan de las Cabezas and so on, Action-Participation-Research initiatives and more (Pangea n.d.; Ciudades para un Futuro más Sostenible n.d.b; Rodríguez-Villasante 2001). In the following section, I discuss the Community Plan of Trinitat

---

4   Similarly, integrated approaches were implemented in Madrid in the 1990s (the Investment Plan) within the framework of the Anti-poverty Plan in eight southern districts. And, while the programme was closed down by the new conservative regional government in 1995, the basic approach was upheld by a grassroots organization (Coordinadora de Movimientos Ciudadanos-Movimiento por la Dignidad del Sur) that in recent years has been key in the development of Community Plans in the municipalities of Villaverde and Usera (Ciudades para un Futuro más Sostenible n.d.c).

Nova in Barcelona, a paradigmatic example of a radical bottom-up initiative for social regeneration in a peripheral and disadvantaged neighbourhood.

Trinitat Nova: Empowered Participation for Social Regeneration

Trinitat Nova is one of the many housing estates built around Barcelona in the 1950s and 1960s as a public sector initiative to accommodate the intense flow of immigrants arriving from rural areas and less prosperous regions of Spain. Located in the north-eastern district of Nou Barris, adjacent to the major ring road of the city, this working-class enclave of little more than 10,000 residents and 55 ha was built in the periphery of the city without basic planning for transport, public spaces, urban services and infrastructure and employing a very poor quality of construction, reflecting a pattern of socio-spatial segregation characteristic of urban development during the Francoist era. Improvements in collective services and infrastructure evolved gradually in response to social pressures from a deeply-rooted and effective neighbourhood association. However, in the late 1970s these gains were rapidly levelled off by the effects of economic restructuring on residents' employment and incomes, which exacerbated problems of severe physical decay of housing stock, demographic loss and aging, spatial segregation and social exclusion, transforming this neighbourhood into one of the most problematic in Barcelona.

The crisis in Trinitat Nova peaked in 1992 when a local housing organization revealed that 600 housing units in the neighbourhood were affected by *aluminosis*,[5] *carbonatosis* and other architectural pathologies that put them at high risk of crumbling. Local authorities responded with a partial restoration and follow-up programme, viewed as inadequate by neighbourhood organizations who demanded total demolition and replacement of all affected buildings. Four years later, the situation had become unsustainable with a growing number of units affected, no comprehensive response from the local and regional administrations, and a declining mobilization capacity on the part of the community. In this context, the Neighbourhood Association, a deeply embedded grassroots organization that had led many of the district's battles for improvement since the 1960s, took the lead against institutional oblivion and further decline.

Starting from a critical reappraisal of its own weakened organizational dynamics[6] and the need to engage new social groups and forms of participation,

---

5   Aluminosis is the deterioration of reinforced concrete structures using aluminous cement, making it more porous and brittle. This type of cement was used extensively in the construction industry during the 1950s and 1960s, because it reduced the time necessary for building.

6   In the early 1990s, the Neighbours Association in Trinitat Nova went through a critical phase as an aging membership and lack of generational renovation or capacity to attract new social groups undermined its organizing potential. This trend was by no means

in 1996, the Association sought outside support from an international community development expert to guide the process of re-engagement and participation towards the development of a Community Plan (http://pangea.org/trinova/documentos/democracia.doc). The community planning process involved a participatory action-research methodology that involved four phases – information and consultation; drafting of proposals; deliberation and decision making; and execution – all of which were facilitated by a community task force (Marchioni 2003). Thus, the process began with a 'diagnosis' of the neighbourhood's situation, its structural components, key problems and basic sectors, involving the maximum number of residents in the production of a shared view of the neighbourhood's problems and solutions. After a year of intense consultation and information gathering from the various community actors, in September 1997 the task force presented the *Community Diagnostic*, a document reflecting the general lines of consensus among local residents, technical and political actors regarding all aspects of neighbourhood life: education, social services, culture, housing, employment and so on. The document identified the key problems for different groups as well as the specific fields for intervention: health, culture, economic and urban planning. But this diagnostic was, above all, a process of collective assessment and proposition as well as implication of residents in the solutions to the problems identified (Blanco and Rebollo 2003).

The process of participatory diagnosis ran parallel to extensive networking dynamics among the various community actors including district associations, technical experts, civil servants and the various public administrations that were crucial in securing the necessary expertise, financial resources and political legitimacy for the Community Plan (Rebollo 2002). Several administrations – the City Council of Barcelona and the Social Affairs Department of the Catalonian government – gave support to the project providing funding and encouraging their workers to get involved in the process. After almost two years of negotiations, in July 1997, the local and regional governments and the Neighbourhood Association signed the *Conveni de Barri*, a neighbourhood contract that provided a stable framework of collaboration among the various stakeholders for the implementation of the Plan.

Implementation of the Community Plan for the period 1997–1999 was firmly grounded on pedagogy of participation, securing residents' access to the necessary resources for self-organization, development of participatory mechanisms, integration of public administrations in the process, integration of technical projects with participatory processes, and a focus on process as open, flexible and in constant evolution. Based on these basic principles, the Community Plan of

---

unique to Trinitat Nova; on the contrary, it was experienced by most social movements. But in Trinitat Nova declining participation and activist fatigue coincided with heightening urban problems and initiatives (transport, water, housing) that had a direct impact on the district and, therefore, demanded significant efforts to negotiate and exert pressure on the part of local residents (Rebollo 2002).

Trinitat Nova established several programmes of social intervention and created a complex structure that enabled the coordination between residents' participation and public administration.

From the start, the Neighbourhood Association of Trinitat Nova has been the leading political actor of the Plan supported by the community task force, a small group of experts in community planning and local activists, accountable only to the association and financed with funds granted by the local and regional administrations but channelled directly through the association. These dynamics were crucial in maintaining the autonomy of the task force and its primary focus of enhancing associational dynamics in the district and the emergence of new neighbourhood organizations.

The Community Plan process has succeeded in fulfilling its three basic objectives: first, to reactivate and strengthen the associative fabric of the district; second, to improve public sector intervention in the district; and third, to contribute to a general improvement of the social, economic and physical conditions in the district. Participation is considered a fundamental neighbourhood asset and a necessary means to enhance the neighbourhood's capacity for social change (Rebollo 2002). Currently, the Community Plan brings together around four hundred people engaged in various programmes, projects and activities generated by the plan and implemented by community actors themselves. This process also reflects a radical change in the way public administrations work in the district including the development of innovative modes of intervention based on integrated approaches to problem solving, multilevel and horizontal coordination and permanent interaction with citizenship (Rebollo 2002).

Over the last few years, community planning has gradually moved away from its original focus on housing, urban planning and infrastructure developments towards an enlarged neighbourhood political agenda that includes education, economic promotion and employment and, more recently, environmental and sustainability concerns. More ambitious programmes such as the *Eco-neighbourhood Trinitat Nova* aimed at the ecological reconversion of the area, are now being discussed, and the mobilization of resources around environmental concerns has become an important component of the neighbourhood's future strategy. Building on the *Trinitat InNova* project, which aimed at taking advantage of extensive demolition and rebuilding to transform Trinitat Nova into an eco-neighbourhood, a series of sectoral sustainability studies were drafted in 2003 to adapt the Master Plan to environmental guidelines. Under the title *Eco-Neighbourhood Trinitat Nova* these studies have been synthesized to form an integrated sustainability plan, and extended to the whole district. The innovative momentum is carefully nurtured and maintained through the very process of participatory dynamics, facilitating the restating and discovery of new social needs and demands.

In sum, Trinitat Nova provides a unique example of socially innovative regeneration grounded on intense neighbourhood organization and citizen participation as well as on the development of new collaborative dynamics between neighbourhood associations and public sector institutions. The radical bottom-up and community-led nature of Trinitat Nova's regeneration strategy is actively sustained through strategic action-oriented deliberation. But, at the same time, collaborative engagement within institutional networks has improved the capacity to access critical technical and financial resources. Empowered participation and innovative social dynamics are simultaneously outcomes and preconditions to the process. Indeed, the example of Trinitat Nova suggests that an active citizenship and a conscious community, capable of articulating collective responses, is a necessary condition for sustained and effective change. Yet, enabling institutions and collaborative dynamics between community residents, planners and political representatives contribute to empowered participation. Finally, participatory local action research has also played an important function providing key structures not only to maintain a strategy focused on problem resolution (housing needs, concrete revealed demands and so on) and involving citizens, government officials and technical experts in the generation of solutions, but also to structure the deliberative development of solutions to these problems.

Concluding Remarks

Neighbourhoods have become centre stage in recent years as an increasingly relevant analytical and policy scale. The current focus on neighbourhoods in Spain as a reconstituted scale of governance is part of the reinvention of government and the renewed emphasis on community participation, partnership and empowerment as a critical means to improve policy delivery and outcomes, promote social inclusion and strengthen local democracy. Community participation plays a key role in the new wave of neighbourhood-based policies allowing for the integration of local knowledge, resources and networks in the policy formulation and implementation process. In disadvantaged neighbourhoods, citizen participation is seen as an antidote to technocratic policy making, relying on the capabilities of residents to identify local needs and priorities and contribute to socially innovative responses that make use of the communities' expertise and lived experience.

However, community participation for the revitalization of deprived neighbourhoods has to be understood in the context of renegotiated responsibilities and powers between the state and civil society. The outcome of this negotiation is highly contextual and depends on the particular conditions and local regimes at a particular time. Thus, either participatory governance can be a simple means to rearticulate a consensus-based regime of efficient policy delivery, self-regulation and transferred responsibility at the neighbourhood scale or, on the contrary, citizen engagement in neighbourhood regeneration can be an end in itself, an opportunity for the self-organization of socially excluded groups to gain control over the aims and priorities of regeneration and for genuine empowerment. In the context of rescaled governance, innovative institutional configurations for participatory policy express the commitment to empower communities in the policy-making and implementation process. However, reification of community participation obscures the crucial distinction between these two trajectories and, as a result, fails to address the conditions necessary for effective, self-determined community involvement.

On the other hand, non-intrumentalized forms of citizen participation play a key role in revealing communities' needs and priorities and empowering communities. But this dynamic must be linked to a process of meaningful communication and confrontation of interests between all relevant political subjects (Healey 1997). Deliberative discussion in the public sphere forms a basis for the construction of common goals. However, unequal power relations and entrenched conflict within communities can undermine severely the possibilities of democratic deliberation, stalling integration and reconciliation of differing values and interests. Strengthening the capacity of residents to participate effectively in and influence the decision-making process is, therefore, an integral part of a participatory urban democracy. And yet, in the context of neoliberal politics, innovative institutional arrangements for community engagement are often designed to promote consensus building around neoliberalizing strategies, initiatives and norms by fostering responsible and domesticated forms of engagement while marginalizing or excluding *tout court* other, more challenging and autonomous forms of involvement.

References

AAVV (1998), *Arquitectura y Urbanismo en Ciudades Históricas* (Madrid: MOPU/UIMP).

Alguacil, J. (2006), 'Barrios desfavorecidos: diagnóstico de la situación española', in Vidal Fernandez (ed.).

Allen, J., Cars, G. and Madanipour, A. (2000), 'Social Exclusion in European Neighbourhoods: processes, experiences and responses' Final Report, Project SOE-CT97-3057 (published online) <http://improving-ser.sti.jrc.it>.

Amin, A. (ed.) (1994), *Post-Fordism: a reader* (Cambridge: Blackwell).

Amin, A. (2005), 'Local Community on Trial', *Economy and Society* 34, 612–33.

Anderson, A. (1998), *La Politique de la Ville* (Paris: Syros).

Arias, F. (ed.) (2000), *La desigualdad urbana en España* (Madrid: Ministerio de Fomento).

Atkinson, R. (2000), 'Combating Social Exclusion in Europe: the new urban policy challenge', *Urban Studies* 37: 5–6.

Atkinson, R. and Kintrea, K. (2001), 'Disentangling Area Effects: evidence from deprived and non-deprived neighbourhoods', *Urban Studies* 38:12, 2277–98.

Bellet, C. (n.d.), 'El Plan Especial y la transformación de la ciudad consolidada' (published online) <http://cervantesvirtual.com/servlet/SirveObras/4687273021>.

Blanco, I. (2005), 'Políticas urbanas de inclusión socio-espacial.Inclusión Socio-Espacial. La experiencia de Barcelona', Paper presented at the X Congreso Internacional del CLAD sobre la Reforma del Estado y de la Administración Pública (Santiago de Chile).

Blanco, I. and Gomá, R. (2003), *Gobiernos locales y redes participativas* (Barcelona: Ariel).

Blanco, I. and Rebollo, O. (2003), 'El Plan Comunitario de la Trinitat Nova (Barcelona): un referente de la planificación participativa local', in Blanco and Gomá (eds).

Busquets, J. (1992), 'Evolución del planeamiento urbanístico en los años ochenta en Barcelona', *Ciudad y Territorio* 93: 31–51.

Campos Venuti, G. (1981), *Urbanismo y Austeridad* (Madrid: Siglo XXI).

Carmon, N. (1997), 'Neighbourhood Regeneration: the state of the art', *Journal of Planning Education and Research* 17: 131–44.

Ciudades para un Futuro más Sostenible (n.d.a), Un ejemplo de participación y renovación urbana: la remodelación de barrios en Madrid (España) [webpage] <http://habitat.aq.upm.es/bpes/onu/bp258.html>.

Ciudades para un Futuro más Sostenible (n.d.b), Programas de acción participativa en diferentes municipios (España) [webpage] <http://habitat.aq.upm.es/bpes/ceh2/bpes36.html>.

Ciudades para un Futuro más Sostenible (n.d.c), Biblioteca CF + S [webpage] <http://habitat.aq.upm.es>.

Couch, C., Fraser, C. and Percy, S. (eds) (2003), *Urban Regeneration in Europe* (London: Blackwell).

Esteban, M., Uhalde, I., Rodríguez, A. and Altuzarra, A. (eds) (2008), *Territorios Inteligentes: Dimensiones y Experiencias Internacionales* (Madrid: Netbiblio).

European Commission (1997), *Towards an Urban Agenda in the European Union* (Brussels [COM (97/197)]).

European Commission (1998), *European Union Framework for Action for Sustainable Urban Development* (Brussels [COM/1998/605]).

Forrest, R. and Kearns, A. (1999), *Joined up Places? Social cohesion and neighbourhood regeneration* (York: Joseph Rowntree Foundation).

Fung, A. and Wright, E.O. (eds) (2003), *Deepening Democracy: institutional innovations in empowered participatory governance* (London: Verso).

Geddes, M. (1997), *Partnership against Poverty and Exclusion? Local regeneration strategies and excluded communities in the UK* (Bristol: Policy Press).

Generalitat di Catalunya (n.d.), Programa de barrios y áreas urbanas de atención especial [webpage] <http://www10.gencat.net/ptop/AppJava/es/arees/ciutat/barris/index.jsp>.

Geyer, H. and Richardson, W. (eds) (2007), *International Handbook on Urban Policy, Part I* (Northampton, MA: Edward Elgar).

Glennerster, H., Lupton, R., Noden, P. and Power, A. (1999), *Poverty, Social Exclusion and Neighbourhood: studying the area bases of social exclusion*, Case Paper 22 (London: Centre for Analysis of Social Exclusion, London School of Economics).

Healey, P. (1997), *Collaborative Planning: shaping places in fragmented societies* (London: Macmillan).

Healey, P. (1998), 'Institutional Theory, Social Exclusion and Governance', in Madanipour et al. (eds).

Hull, A. (2001), 'Neighbourhood Renewal: a toolkit for regeneration', *GeoJournal* 51, 301–10.

Jacquier, C. (1990), *Voyage dans Dix Quartiers Européens en Crise* (Paris: L'Harmattan).

Jessop, B. (1994), 'Post-Fordism and the State', in Amin (ed.).

Kearns, A. and Paddison, R. (2000) 'New Challenges for Urban Governance', *Urban Studies* 37:5–6, 845–50.

Kearns, A. and Parkinson, M. (2001), 'The Significance of Neighbourhood', *Urban Studies* 38:12, 2103–10.

Keating, D., Krumholz, N. and Star, P. (eds) (1996), *Revitalizing Urban Neighbourhoods* (Lawrence, KS: Kansas University Press).

Keating, D. and Smith, J. (1996), 'Neighborhoods in Transition', in Keating, Krumholz and Star (eds).

Leal, J. (1989), 'La sociología y el urbanismo en los últimos diez años', *Ciudad y Territorio* 81:2/3–4, 39–43.

Leitner, H., Peck, J. and Sheppard, E.S. (eds) (2007), *Contesting Neoliberalism: urban frontiers* (New York: Guilford Press).

Lipietz, A. (1989), *La société en sablier: le partage du travail contre la déchirure sociale* (Paris: La Découverte).

Madanipour, A., Cars, G. and Allen, J. (eds) (1998), *Social Exclusion in European Cities* (London: Jessica Kingsley).

Marchioni, M. (2003), 'Democracia participativa y crisis de la política. La experiencia de los planes comunitarios' in Jornadas sobre democracia participativa, Sarriko, 26–27 June 2003.

Marcuse, P. and van Kempen, R. (eds) (2000), *Globalizing Cities: a new spatial order?* (Cambridge: Blackwell).

Martens, A. and Vervaeke, M. (eds) (1997), *La polarisation sociale des villes européennes* (Paris: Anthropos).

Mingione, E. (1996), *Urban Poverty and the Underclass* (Cambridge, MA: Blackwell).

Morrison, N. (2003), 'Neighbourhoods and Social Cohesion: experiences from Europe', *International Planning Studies* 8:2, 115–38.

Moulaert, F. et al. (2000), *Globalization and Integrated Area Development* (Oxford: Oxford University Press).

Moulaert, F., Alaez Aller, R., Cooke, Ph., Courlet, Cl., Häusserman, A. and da Rosa Pires, A. (1990), *Integrated Area Development and Efficacy of Local Action: feasibility study for the European Commission* (Brussels: DG V).

Moulaert, F., Delladetsima, P. Delvainquière, J.-C. et al. (1994), *Local Development Strategies in Economically Disintegrated Areas: a pro-active strategy against poverty in the European Community*, Report for the EC DG Research (Lille: IFRESI-CNRS).

Moulaert, F., Martinelli, F., González, S. and Swyngedouw, E. (2005), 'Towards Alternative Model(s) of Local Innovation', *Urban Studies* 42:11, 1969–90.

Moulaert, F., Martinelli, F., Gonzalez, S. and Swyngedouw, E. (2007), 'Introduction: Social Innovation and Governance in European Cities. Urban development between path-dependency and radical innovation', *European Urban and Regional Studies* 14:3, 195–209.

Moulaert, F., Morlicchio, E. and Cavola, L. (2007), 'Social Exclusion and Urban Policy in European Cities: combining "Northern" and "Southern" European perspectives', in Geyer and Richardson (eds).

Moulaert, F., Rodríguez, A. and Swyngedouw, E. (eds) (2003), *The Globalized City: economic restructuring and social polarization in the European city* (Oxford: Oxford University Press).

Murie, A. and Musterd, S. (2004), 'Social Exclusion and Opportunity Structures in European Cities', *Urban Studies* 41:8, 1441–59.

OECD (1998), *Integrating Distressed Urban Areas* (Paris: OECD).

Pacione, M. (ed.) (1997), *Britain's Cities: geographies of division in urban Britain* (Harlow: Longman).

Pangea (n.d.), project website <http://pangea.org>.

Parkinson, M. (1998), *Combating Social Exclusion: lessons from area-based programmes in Europe* (Bristol: Policy Press).

Peck, J. and Tickell, A. (2002), 'Neoliberalizing Space', *Antipode* 34:3, 380–404.

Pol, F. (1988), 'La recuperación de los centros históricos en España', in AAVV.

Priemus, H. (2005), 'Urban Renewal, Neighbourhood Revitalization and the Role of Housing Associations: Dutch experience', Paper presented at the National Policy Forum on Neighbourhood Revitalization (Ottawa, 25 October).

Raco, M. (2000), 'Assessing Community Participation in Local Economic Development: lessons for the new urban policy', *Political Geography* 19, 573–99.

Raco, M. and Imrie, R. (2003), 'Towards an Urban Renaissance? Sustainable communities, New Labour and urban policy', Paper presented at Regional Studies Association Conference Reinventing the Regions (Pisa, 12–15 April).

Rebollo, O. (2002), 'Metodologías y Prácticas Transformadoras. El Plan Comunitario de Trinitat Nova' (published online) <http://pangea.org/trinova/documentos/ponenciaoscar.doc>.

Rodríguez, A. and Abramo, P. (2008), 'Innovación Institucional, Gobernanza Democrática y Desarrollo Urbano. El Caso de Porto Alegre, Brasil', in Esteban et al. (eds).

Rodríguez, A., Moulaert, F. and Swyngedouw, E. (2001), 'Nuevas politicas urbanas para la revitalizacion de las ciudades en Europa', *Ciudad y Territorio Estudios Territoriales* 33:129, 409–24.

Rodríguez, A. and Vicario, L. (2005), 'Innovación, Competitividad y Regeneración Urbana: los espacios retóricos de la "ciudad creativa" en el nuevo Bilbao', *Ekonomiaz* 58:1, 262–95.

Rodriguez-Villasante, T. (2001), *La complejidad y los talleres de creatividad social* (Madrid: El Viejo Topo).

SINGOCOM (2005), *Final Report (2005) Social Innovation, Governance and Community Building*, Contract nr: HPSE-CT2001-00070.

Swyngedouw, E. (2005), 'Governance Innovation and the Citizen: the Janus face of governance-beyond-the-state', *Urban Studies* 42:11, 1991–2006.

Swyngedouw, E., Moulaert, F. and Rodríguez, A. (2002), 'Neoliberal Urbanization in Europe: large-scale urban development projects and the new urban policy', *Antipode* 34:3, 542–77.

Terán, F. (1999), *Historia del urbanismo en España III. Siglos XIX y XX* (Madrid: Cátedra).

Van den Berg, L., Braun, E. and Van der Meer, J. (1998), 'National Urban Policy Responses in the European Union: towards a European urban policy?' <http://www.ersa.org/ersaconfs/ersa98/papers/435.pdf>.

Vidal Fernández, F. (ed.) (2006), *V Informe FUHEM de Políticas Sociales: la exclusión social y el estado del bienestar en España* (Madrid: FUHEM).

Wilson, W.J. (1997), *When Work Disappears: the world of the new urban poor* (New York: Alfred A. Knopf).

Chapter 6

How Socially Innovative is Migrant Entrepreneurship? A Case Study of Berlin

Felicitas Hillmann

Introduction

By now, a variety of immigrant groups in Germany have become strongly represented in self-employment, with some German cities reporting that foreigners are responsible for at least a quarter of all new business registrations. This phenomenon points to broader changes in the urban economic setting. Furthermore, in recent years 'ethnic economies' – basic forms of migrant entrepreneurship – have been increasingly welcomed by some city administrations as important tools for urban development itself. Germany denied being an immigration country for many years (among many see: Hillmann 2007; Moulaert and Philips 1982; Nikolinakos 1973) and did not favour ethnic entrepreneurship actively – so why would it favour it now?

While in a first phase the initiative to push 'ethnic economies' as one possible form of economic organization came from migrant groups themselves, increasingly interest groups and institutions from 'the outside' have appeared on the scene and are trying to 'organize the field'. In addition, EU-financed urban programmes relate much more to existing immigrant realities at the community level than they did during the 1990s, and now actively involve the immigrant population as strategic partners in their activities – even though there is no deliberate strategy embodying a 'multicultural policy' within cities and urban communities. Strategies that initially focused on bottom-up economic development now emphasize urban development and planning. Clearly 'bottom-up economic activities' are no longer considered a common good in urban regeneration strategies, which today are geared more toward the improvement of market feasibility and economic investment (Gerometta, Häussermann and Longo 2005).

Until recently 'ethnic economies' have been conceptualized in the literature as predominantly economic strategies through which migrants are able either to translate their cultural resources into economic ones, or to escape economic marginalization. This chapter aims to introduce 'ethnic economies' as possible key actors in the process of social innovation, hence as a strategy of social innovation itself, and claims that forms of ethnic entrepreneurship might be interpreted as part of what is labelled the 'social economy'. It is argued that development strategies from below and from above coexist in the field of ethnic entrepreneurship and that this is a new response to marginalization for certain social groups at risk. Here, the role and function of 'ethnic economies' transcend economic rationalities in many respects – for example because they often exert a very strong social function, sometimes more influential than their economic function. According to Hillier, Moulaert and Nussbaumer (2004) social innovation takes place when initiatives of development from below, from the actors on the ground, meet (and occasionally merge) with institutional initiatives intended to foster development from above, and when the new forms of social organization that are created in this way are persistent over time. Moulaert et al. (2007) stress three dimensions when defining social innovation: the satisfaction of human needs (which are 'not yet' or 'no longer' perceived as important by the state or by the market and thus not satisfied); changes in social relations (and in particular an increase in the level of participation); increasing socio-political capability and access to resources (empowerment).

This chapter links the concept of ethnic entrepreneurship with that of social innovation. In its first section it develops the concept of ethnic entrepreneurship as it is treated in the existing body of scientific literature – especially in the United States, but with noticeable differences on the European scene. None of the authors interpret ethnic entrepreneurship in terms of social innovation, and only rarely do they relate their analyses to changes in urban governance. Ethnic entrepreneurship is seen either as an economic strategy, or as a step towards integration or assimilation, sometimes linked to social mobility (Kesteloot and Mistiaen 1997). The second section presents the case of Berlin. The focus here is two highly important immigrant groups in Berlin in terms of numbers and in terms of migration structure: the Turkish and the Vietnamese ethnic populations. After portraying the situation of ethnic economies, the chapter addresses the institutional setting that is starting only recently to evoke and create 'ethnicity' as an aspect of urban development.

From the United States to Europe: The Theoretical Underpinnings of Ethnic Entrepreneurship

Until recently the bulk of literature on ethnic entrepreneurship has come from the United States, often serving as an implicit reference for European research. In the United States, where the whole process of nation building had been conceptualized following the melting pot ideology that merged successive generations of immigrants into one big 'America', the racial unrest of the 1960s brought race relations back to the centre of sociology. Sociologists turned away from the idea of the 'melting pot' as a metaphor for assimilation. The dominant notion for many years then saw the United States as an ethnopluralistic society in which ethnic minorities existed side by side without mixing – the metaphor used here was that of the 'salad bowl'. From the 1970s middlemen minority theories gave special emphasis to cultural features of ethnic minorities and their predisposition towards certain types of enterprises (cf. Bonacich 1973, relying on the Simmeliarian idea of the stranger). In the tradition of social ecology, the human–ecological model of succession was applied to 'black' entrepreneurs in 'white' residential areas (cf. Aldrich and Reiss 1976); culturally influenced rotating credit systems were thematized to explain the success of certain ethnic groups or the lack of 'black' success (Light 1972).

Europe has never faced such clear cut 'racial' settings, and also the idea of the 'self-made man' is of far less importance due to often highly regulated and historically rooted segmented labour markets – so we might expect that this diversity is also mirrored in the literature.

Among European countries, Britain was first to discover that ethnic economies could work as important components of urban economies. The political hope was to revitalize run-down inner cities in Britain, with ethnic economies playing a role (Haberfellner 2000). Unlike Britain, other European states had rigid legal regulations hindering immigrants from starting their own businesses. Ethnic enterprise, which in many European countries is mainly located in the informal sector of the labour market, is therefore likely to be discussed in close connection with processes of social exclusion (Samers 1998). It should be remembered in this respect that immigrants and minorities in Europe were for many years concentrated in the least attractive segments of the labour market and on the margins of the formal sector (see Mingione 1999). Further, as Kesteloot and Mistiaen (1997, 244) point out, in the early literature on ethnic entrepreneurship, the close relationship of ethnic entrepreneurship and informal practices was an important theme.

More recently, intra-European comparisons have underlined the influence of institutional frameworks on ethnic businesses in Europe and have relegated cultural factors to the background as minor influences. There has also been an attempt to provide a theoretical explanation for the considerable differences between ethnic businesses – sometimes within the same ethnic group – in various European cities. Ethnic enterprise is analysed in the context of 'migration regimes' and the focus is on the socioeconomic affiliation of the ethnic entrepreneurs rather than on individual enterprises (Kloosterman and Rath 2003; Kloosterman et al. 1999 for the concept of 'mixed embeddedness').

Another, even more sceptical reading pronounces the exploitative nature of this kind of business. Jones, Ram and Edwards (2006) state that with ongoing globalization the (self-)exploitation of immigrant labour grew too – as part of the worldwide race to the bottom. A fundamental problem of both US and European research on ethnic economies is that in many countries little statistical information is available on the situation of similar native-run businesses. Consequently there is always the danger of an 'ethnic earmarking' of certain economic activities. Another problem is that an 'ethnic approach' tends to categorize all immigrants primarily within 'ethnic groups', irrespective of their gender, class or professional background, and thus overlooks the importance of social stratification. A standard definition of the 'ethnic economy' has yet to come. It is generally understood as a certain type of spatial cluster of ethnic enterprises (that is, those owned by a non-indigenous population group). The literature emphasizes five particular attributes of these enterprises: first, the horizontal and vertical networking between them (networking between enterprises of the same type and among production lines); second, their reliance primarily on co-ethnic employees and suppliers; third, their tendency to address their respective ethnic community as clientele; fourth, the continuous involvement of family members; and fifth, the inclination to identify with their own ethnic group (cf. Hillmann 1998; IMF 2005).

The Berlin Example

Berlin is a city with a visible immigrant population. Of its 3.4 million inhabitants, 795,000 have a migrant background – persons originating from immigrant families or who are migrants themselves – equivalent to 23.38 per cent of the population. Certain neighbourhoods have 40 per cent or more immigrants among their registered residents: these are mostly poorer parts of Berlin, like Neukölln, Kreuzberg and Mitte (Wedding) – all located in inner-city districts. Three factors are relevant to grasp the impact of 'ethnic economies' and their relation to social innovation: the general situation in the labour market – indicating opportunities and barriers for the immigrant population; the demography of business creation and closure; the role of institutions as facilitators of economic action as well as their place in the promotion of 'ethnic' festivals and events.

Labour Market Performance

The Berlin labour market since the late 1990s is considered a difficult one – especially for those not holding a German passport (see Figure 6.1). According to micro-census data, unemployment among the German population in the capital city was as high as 17.1 per cent in 2005 (1995: 13.3 per cent), much higher than in most other *Länder* [states] of the country. Among the foreign population, unemployment rose in ten years to 32.8 per cent (1995: 28.3 per cent). Among the Turkish population unemployment was at its highest peak ever at 46.5 per cent (1995: 33.8 per cent) (see Figure 6.1). After reunification, Berlin suffered sluggish economic development – many firms formerly subsidized by the Western federal government turned away from Berlin; few new firms could be attracted. The process of deindustrialization had a very strong impact on precisely those districts in which the concentration of working-class and immigrant populations was traditionally high. Large numbers of unemployed, as well as a high drop-out ratio among (young) foreigners,[1] are now among the most salient problems in those inner-city districts.

---

1 German statistics on migration and integration distinguish between foreigners (those with no German passport); non-German pupils (only in school-statistics) and for some years also unemployment among nationalities.

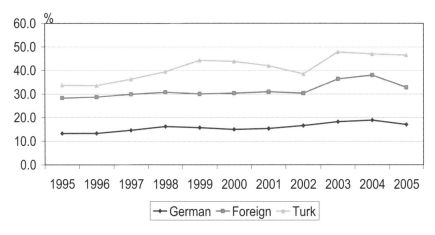

Figure 6.1   **Unemployment rates of the German, foreign and Turkish population in Berlin, 1995–2005**

Self-employment became one of the most favoured strategies to escape marginalization on the labour market, and today in Berlin the proportion of foreigners who are self-employed is higher than that of the native German population without a migratory background.

Corporate Demography: Business Registrations and Cancellations

In 2005, the total percentage of business registrations in Berlin that came from foreigners was as high as 29.5. When considering business registrations and cancellations – the only reliable indicator on ethnic entrepreneurship – it becomes clear that those conducted by the Turkish and Vietnamese, the two most prominent groups, have tended to be stable over the past 15 years (Figure 6.2).[2] This leads to the question of why the role of ethnic economies changed at this time, even though business dynamics had not changed much since the early 1990s.

Turkish and Vietnamese migrants are amongst the most pioneering when it comes to ethnic entrepreneurship in Berlin. While Turkish entrepreneurship is mainly bound to the former western part of the city (where the Turkish population lived before reunification), Vietnamese entrepreneurs are located much more in the eastern part of the town (formerly the capital city of the GDR). While the

---

2   The strong rise of Polish entrepreneurship since 2003 reflects Poland's geographical situation as a neighbouring country and the consequence of its admission to the EU. In contrast to the Turkish and Vietnamese migrant entrepreneurs, we might consider the Polish case as exceptional since it relies on a community which has the possibility to travel and commute and therefore, is much more flexible in its migratory patterns than migrants from more distant and non-EU countries (Morokvasic 1994: 1975).

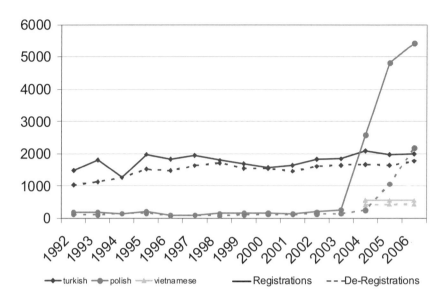

**Figure 6.2   Turkish, Polish and Vietnamese business registrations
and cancellations in Berlin, 1992–2006**

Turkish community has been well documented, research on the socio-economic
integration of the Vietnamese has only just begun. The data presented here stem
from my own and students' field work, official statistics and related publications.
It highlights common ground and differences among the two groups.

Turkish entrepreneurship as a strategy from below   At the local level, the Turkish
'ethnic economy' in Berlin is by now well established, as it is in most urban
agglomerations in Germany (Scholz 1996; Goldberg and Sen 1997; Leicht et al.
2005). In 2004, there was also a well established network in Berlin of nearly forty
Turkish and Kurdish associations that take care of the interests of this immigrant
group and also promote self-employment among its members. Recent studies on the
Turkish ethnic economy in Berlin confirm the findings that self-employment among
Turks tends to concentrate in certain areas and niches of the economy and that they
relate to each other as harsh competitors for a small market. Turkish entrepreneurs
have spread everywhere in the western part of town and the concentration of Turkish
businesses is now greater in districts where the percentage of residential Turkish
immigrants is low than in those districts where it is high (Pütz 2004). However, only
a few Turkish entrepreneurs have dared to open up businesses in the eastern parts
of the town. Examining the clientele of the Turkish businesses, Pütz' findings show
that the proportion of clients of Turkish origin is still high, even if the enterprise
is located in a predominantly 'German' setting. An unpublished field survey in
2006 in the neighbourhood of Neukölln, Rixdorf, showed that 19 of a total of 43

enterprises were run by migrants of Turkish origin, four by Vietnamese and four by other nationalities, while 16 were run by Germans. In this neighbourhood German enterprises are mostly active in services, while retail, trade and craft are not as important as for their Turkish counterparts. The entrepreneurship of the Turkish community today shows a mixture of established and well organized entrepreneurs as well as a stratum of marginalized and precarious ones, especially in the sectors with low entrance barriers. Wilpert (2003) argues that the Döner was 'invented in Berlin' and Rudolph and Hillmann (1997) claim that the inclusion of the Döner in the traditional food sector in Berlin is part of a broader ethnicization of this segment of the economy. Today Döner snack bars are present in nearly all parts of Berlin. Simultaneously, a broad network of Turkish (business) associations has developed, representing the interests of the Turkish community in the media and in public life.

Vietnamese entrepreneurship as a strategy from below Confronted with problems of unemployment and unable to find new jobs on the Berlin labour market, many Vietnamese immigrants try to establish their own enterprises mainly in the retail food, flower and grocery sectors. Indeed, in these sectors in 2003 we find widespread Vietnamese activities in most parts of Berlin. Whereas this kind of business prevails in the inner districts, in the outskirts trade centres began to be established in the mid 1990s. More than fifty traders with about a hundred employees, four translation bureaus, and a communal consultancy office were located in one of these centres. By 2007, the first Asian trade centre made its way to the inner city, indicating a growing organizational initiative among some members of the Vietnamese community.

The majority of Vietnamese businesses are small and they are based on the integration and the exploitation of all potential labour power, often in an informal way involving relatives, and especially wives, husbands and children. Only in rare cases are more than four or five persons found working together. The existence of the Vietnamese ethnic economy is the result of limited access to the formal labour market and limited options to secure an income for the majority of the migrants (Schweizer 2004; Hillmann 2005). In the early 1990s, the Vietnamese ethnic economy became established as an enduring part of the local economy. In 2003, the Berlin Chamber of Industry and Commerce (IHK; *Industrie und Handelskammer*) indicated that nearly 10 per cent of the officially registered Vietnamese population in Berlin is registered as self-employed (983 Vietnamese businesses registered). Over the past years a widespread network of loosely connected associations has also developed in this community. Today about twenty associations and institutions concentrating on Vietnamese immigrants work in Berlin and provide some form of institutional representation/partnership for requests from the Berlin Senate. One of the strategies of the Vietnamese in Berlin was the expansion of their businesses into subway stations and onto the territory of public transport – especially at the crossing points of connecting lines; in a way this demonstrated the strategy of 'going underground' by utilizing all possible spaces available to them. Transit points, such as train stations, constituted a new and privileged market for small low-cost services such as floristry. Field work in Berlin showed that on two main metrolines (U7 running from east to west and U9 running from north to south) nine of a total of 19 flower shops were run by Vietnamese immigrants. An additional feature is the presence of ambulant flower shops at various public transport stations.

However, from a merely statistical viewpoint (number of business registrations and cancellations), the data presented above indicates stagnation among the business activities of Turkish and Vietnamese migrants. The analysis of the social and economic situation of entrepreneurs in the field shows that most of them use their business as a survival strategy, that is, as probably the only remaining option to make a living. So the question remains as to why ethnic economies over past years have still been encouraged by policy makers as tools for urban development.

Attempts towards Institutionalizing Ethnic Entrepreneurship in the Urban Structure: Strategies from Above

I now turn to an examination on the local level of the roles of institutional governance and civil society in particular, in an attempt to understand how ethnic entrepreneurship became a tool for urban development in those districts in which the migrants live. Two fields of interest arise: first, the institutions represented by the Berlin Senate, which shifted its focus from the purely economic to the socioeconomic dimension in terms of area development; and second, the tendency to integrate 'ethnic' activities into a wider process of what might be called 'festivalization and marketization' of ethnic neighbourhoods. As stated above, the institutional setting has changed over the past ten years: on one hand, there is a denser and more established network of immigrant associations; on the other, after long years of neglect of Germany as an immigrant country, a more deliberate policy towards immigrants has been introduced – generating new expectations towards immigrant entrepreneurship.

The Institutional Dimension

For a long time, ethnic entrepreneurship was discouraged by public and private institutions, because of strict and obscure regulations. Among the first institutions to discover the potential resources of ethnic economies were the IHK (Chamber of Industry and Commerce), Turkish lobby groups and business associations. Originally these were interested in the possibility of finding enterprises which would offer vocational training to young people. In the meantime, several official programmes had been launched for vocational training in enterprises run by foreigners. The local administration became active in enhancing migrant entrepreneurship and took notice of existing EU programmes such as EQUAL, among others. Furthermore, in accordance with Berlin's plans to sharpen its profile as a European business location, start-ups and business activity by foreigners were, and continue to be, increasingly considered a significant factor of local economic life. Of interest here are not only investment decisions by large companies, but also start-ups by non-Germans resident in Berlin.[3] The local administration has made available information on entrepreneurship in various languages, offered consulting and training in the field of self-employment, and sought possibilities to offer microfinance credit to migrants.

The report of the Schader Foundation (2005) confirms the growing importance of ethnic entrepreneurship for urban development. The trend to appreciate the presence of ethnic entrepreneurship basically in economic terms was reinforced with the implementation of several EU-financed projects. Certain districts had become laboratories of urban development and social revitalization, but until the middle of the 1990s, ethnic entrepreneurship played only a minor role in this process (see Seidel 2005). A deliberate 'multicultural' policy was still missing in 2005. Nor did commissioners for the integration of foreigners always exist at the district level – even though there was a visible migrant population. In Neukölln, a district with a large visible migrant population, one commissioner explained in 2005: 'Many reject the idea that we are an immigrant district and many express openly their fear that certain activities would lead to *Überfremdung* [alienation]'[4]. She admitted that there was little coordination between the different scales of action and no common integration model at all. Still, from the institutional side there was clearly an attempt to integrate migrants into the activities of the *Quartiersmanagement* [district management]. The local managers, often immigrants themselves, were on the whole able to find adequate solutions to local problems, to ease social tensions and to stimulate alternative, non-profit oriented forms of community life; however, they could offer little assistance in the case of economic, more structurally rooted, problems. In the future, community managers could try even harder to integrate the immigrant population into their projects and to foster self-employment among migrants. In the evaluation of the programme on district management the Berlin Senate states:

> Priority is given to the increased involvement of foreign inhabitants in the community's development. Many of the neighbourhoods in which the programme is run show high percentages of immigrants and of different cultures. It is true that in the past these groups have been addressed, but given the importance of integration for the successful development of these neighbourhoods these endeavours are to be pushed even more. There is a need for specially trained persons and special projects that could make better use of the resources and potentials of these groups. Immigrants have to be involved in the projects and everybody should recognize that. (Abgeordnetenhaus Berlin 2004, 12)

---

3   This information was provided by officials at Berlin's Senate Department of Economics and Technology.

4   English translation of interviews by the author.

Local authorities of the city administration are now making big efforts to include migrants as social innovators in projects for community development. And in 2007, after two years of preparation, a group of immigrants from various countries succeeded in establishing an association which is supported by the administrative body and which will serve as a platform for immigrant entrepreneurs (*Initiative Ethnische Ökonomien*).

Further, the city administration hopes to enhance the status of the diverse cultures the migrants bring into the district – for example through artists who function as 'urban pioneers', using spaces no longer rented. 'Culture' and services are now seen as important tools for urban development not only by the local administration, but also by many artists that live and work in those districts precisely because rents are low there. A bundle of initiatives was launched, such as awards for the activities of new foundations like the *Neuköllner Bürgerstiftung* (Civic Foundation of Neukölln) which is supported by migrant communities and is an outcome of, among other things, the networking of the local administration and associations in Neukölln. Another long-running initiative was the 'international lunch' project, which weekly offered a lunch based on ethnic cuisine for low income groups in a state-run community centre. A mixture of citizens of all ages and backgrounds came to join – about seventy to eighty each time. Further, a festival called *48-hours Neukölln* relies very much on the input of migrant groups and has been successful for about eight years. Future plans to make the Neukölln district more attractive include a Hindu Temple and a Bazaar with shops – a sign of the high social and symbolic value of migrant activities in the district. Construction work for the temple started in 2007.

However, partners in the local administration were sceptical about whether these highly successful initiatives could ever survive without direct or indirect sponsoring through state-subsidized work programmes. The local administration is well aware that such activities cannot resolve the structural problems that affect the district, but could help to keep social tensions within the district under better control. Asked about what prospects there might be for the district, the interviewed partners suggested '*tourism*', creating an *intercultural culture in the local administration* (which sees migrants not as a threat but as a potential), and improving *the quality of the schools* in the district – where German-speaking children are now scarcely to be found.

Festivalization and Marketization

The festivalization of urban policies through ethnic parades, as described by Zukin for New York City (1995), is now an important pillar of Berlin's tourism industry. In their first phase, the main actors in Berlin were the migrants themselves and their interest groups – organizing an event for the locals. The local administration and NGOs in the field of social work did not immediately realize the importance of such an event for the Neukölln district. Today, parades and festivals are clearly linked to issues of local and community development and to the existing ethnic economies

– the organizers of those events are more than likely to support this kind of activity. As one privileged observer in the local administration of Neukölln, reflects:

> When I first heard about the idea to have a carnival parade in Berlin (in May or June!) in 1996, I laughed and I judged it an absurd idea. Then the organizers moved to Kreuzberg (which is the district nearby) and now, as the festival became a huge success, we regret that we were so stupid to let them go. We just did not realize the potential of the thing. ... We underestimated the dimensions of the project.

The parade, which first started as a small project launched by migrant groups in the Neukölln area, subsequently became a commercialized event in neighbouring Kreuzberg. Starting as a parade and a street party with initially 50,000 visitors, the parade today attracts about 1.5m visitors and is the most popular parade in the town; the concept was then adopted by many other German cities.

Associated with this trend towards an ethnicization of cultural activities is the rise of 'ethnomarketing' seeking to connect the ethnic entrepreneurs and to serve as an advertising platform. Since the autumn of 2002, the magazine *Ethnotrade* has been published about three times a year, with a circulation of 20,000 and with generally about fifty pages in each issue. The magazine is part of a set of marketing strategies for ethnic entrepreneurship: the responsible editor, Bernhard Heider, also organizes a fair, a telephone card business and an award for the most successful entrepreneur of the year, and considers the variety of ethnocultures as stimulants for marketing strategies.

Conclusions: Migrants as Local Actors in the Urban Milieu – Social Innovation

As discussed earlier, the concept of ethnic economy came from the United States and was often naively transplanted into the European situation. Recently, more 'European' approaches, for example that of 'mixed embeddedness', call for more differentiated studies, taking into account structural, local and cultural variables. The Berlin case presented here shows that ethnic entrepreneurship has a strong presence in Berlin, owing initially to a growing marginalization of migrants on the local labour market. Having said that, statistics also show that there is stagnation in business registrations and de-registrations. Nevertheless, a heterogeneous alliance of interest groups, local administration and migrant associations has worked successfully to advertise the concept of ethnic entrepreneurship in the past five years. This might be considered a new form of governance, not experienced previously in the city. The local administration and the *Quartiersmanagement*, along with various initiatives sponsored by the EU, have explicitly recognized the importance of the migrant population amongst other economically marginalized groups, such as artists, as innovation agents and urban pioneers for deprived areas and are willing to enhance this 'ethnic or immigrant' aspect of community life even more. Several initiatives that combine commercial and social goals have worked out well, especially when migrant associations have been involved. However, most projects tend to dissolve when no longer subsidized by the local administration, raising questions about their sustainability. Recently, bigger commercial players have gained ground: ethnic journals and ethnic fairs use 'ethnicity' as a marketing strategy.

In a nutshell: there is an ambivalent role for ethnic entrepreneurship and ethnicity in urban development. For most immigrants, ethnic entrepreneurship works as a survival strategy; only for a few did it become a successful business strategy. Nevertheless, from the local administration's perspective, migrant entrepreneurship is increasingly treated as a tool for local development and is seen as a mainstay of more substantial action to help stabilize socially marginalized neighbourhoods. It does, after all, possess a high potential for producing social innovation and, when followed by coordinated actions, for enhancing community development.

Regarding ethnic economies, there has been a clear change in terms of the agenda of formal local administration, as well as among the actors themselves, all of which refers to a process of social innovation that brought 'ethnicity', as one element of urban diversity, into mainstream society. The emergence of development strategies from below and from above is a new approach which transcends purely economic rationalities in many respects (for example, through the marketization of 'ethnicity' and 'culture') and refers to the establishment of a form of social economy. Ethnic entrepreneurship still serves the satisfaction of human needs not sufficiently met by the (labour) market or the state, increased participation for some local actors, an empowerment of previously marginalized people; consequently all three dimensions (outlined in the introduction) that might characterize 'social innovation' are present. The question remains as to whether these changes lead to a better and lasting inclusion of ethnic groups and individuals in the wider society.

References

Abgeordnetenhaus Berlin (2004), *Schlussfolgerungen aus dem Evaluierungsbericht zum Quartiersmanagement und künftige Programmumsetzung*. Drucksache 15/2740, 6 April 2004.

Aldrich, H. and Reiss, J. (1976), 'Continuities in the Study of Ecological Succession: changes in the race composition of neighborhoods and their businesses', *American Journal of Sociology* 81:4, 845–67.

Bonacich, E. (1973), 'A Theory of Middleman Minorities', *American Sociological Review* 38, 583–94.

Gerometta, J., Häussermann, H. and Longo, G. (2005), 'Social Innovation and Civil Society in Urban Governance: strategies for an inclusive city', *Urban Studies* 42:11, 2007–21.

Goldberg, A. and Sen, F. (1997), 'Türkische Unternehmer in Deutschland. Wirtschaftliche Aktivitäten einer Einwanderungsgesellschaft in einem komplexen Wirtschaftssystem', *Leviathan Sonderheft* 17, 63–84.

Haberfellner, R. (2000), 'Ethnische Ökonomien als Forschungsgegenstand der Sozialwissenschaften', *SWS-Rundschau* 40:1, 43–61.

Hillier, J., Moulaert, F. and Nussbaumer, J. (2004), 'Trois essais sur le rôle de l'innovation sociale dans le développement local', *Géographie, Economie, Sociétés* 6:2, 129–52.

Hillmann, F. (1998), *Türkische Unternehmerinnen und Beschäftigte im Berliner ethnischen Gewerbe*, Discussion Paper FS I – 107 (Berlin: Wissenschaftszentrum Berlin).

Hillmann, F. (2005), 'Riders on the Storm: Vietnamese in Germany's two migration systems', in Spaan et al. (eds).

Hillmann, F. (2007), *Migration als räumliche Definitionsmacht? Beiträge zu einer neuen Geographie der Migration in Europa* (Stuttgart: Steiner Verlag).

IMF (Institut für Mittelstandsforschung) (2005), *Die Bedeutung der ethnischen Ökonomie in Deutschland. Push- und Pullfaktoren für Unternehmensgründungen ausländischer und ausländischstämmiger Mitbürger*. Studie im Auftrag des Bundesministeriums für Wirtschaft und Arbeit (Universität Mannheim).

Jones, T., Ram, M. and Edwards, P. (2006), 'Ethnic Minority Business and the Employment of Illegal Migrants', *Entrepreneurship and Regional Development* 18, 133–50.

Kesteloot, C. and Mistiaen, P. (1997), 'From Ethnic Minority Niche to Assimilation: Turkish restaurants in Brussels', *Area* 29:4, 325–34.

Kloosterman, R. and Rath, J. (eds) (2003), *Immigrant Entrepreneurs: venturing abroad in the age of globalization* (Oxford: Berg).

Kloosterman, R., van der Leun, J. and Rath, J. (1999), 'Mixed Embeddedness, Migrant Entrepreneurship and Informal Economic Activities', *International Journal of Urban and Regional Research* 23:2, 253–67.

Leicht, R. et al. (2005), *Die Bedeutung der ethnischen Ökonomien in Deutschland* (Mannheim: Institut für Mittelstandsforschung).

Light, I. (1972), *Ethnic Enterprise in America* (Berkeley, CA: University of California).

Mingione, E. (1999), 'Introduction: immigrants and the informal economy in European cities', *International Journal of Urban and Regional Research* 23:2, 209–11.

Morokvasic, M. (1994), 'Pendeln statt Auswandern. Das Beispiel der Polen', in Morokvasic and Rudolph (eds).

Morokvasic, M. and Rudolph, H. (eds) (1994), *Wanderungsraum Europa. Menschen und Grenzen in Bewegung* (Berlin: Sigma).

Moulaert, F., Martinelli, F., Gonzalez, S. and Swyngedouw, E. (2007), 'Introduction: Social Innovation and Governance in European Cities. Urban development between path-dependency and radical innovation', *European Urban and Regional Studies* 14:3, 195–209.

Moulaert, F. and Nussbaumer, J. (2005), 'The Social Region: beyond the territorial dynamics of the learning economy', *European Urban and Regional studies* 12:1, 81–100.

Moulaert, F. and Philips, F. (1982), 'The Employment of Migrant Workers in the Belgian Economy: a structural phenomenon', *Environment and Planning A* 14:7, 857–67.

Nikolinakos, M. (1973), *Politische Ökonomie der Gastarbeiterfrage* (Hamburg: Rowohlt).

Pütz, R. (2004), *Migration und Transkulturalität* (Bielefeld: Transcript Verlag).

Rudolph, H. and Hillmann, F. (1997), 'Döner contra Bulette – Döner und Bulette: Berliner türkischer Herkunft als Arbeitskräfte und Unternehmer im Nahrungsgütersektor', *Leviathan Sonderheft* 17, 106–20.

Samers, M. (1998), 'Immigration, "Ethnic Minorities", and "Social Exclusion" in the European Union: a critical perspective', *Geoforum* 29:2, 123–44.

Schader Foundation (eds) (2005), *Zuwanderer in der Stadt: Expertisen zum Projekt* (Darmstadt: Schader Stiftung).

Scholz, F. (ed.) (1996), *Ethnisches Gewerbe in Berlin Kreuzberg*, Occasional Paper (Berlin: Freie Universität Berlin Press).

Schweizer, J. (2004), 'The Vietnamese Ethnic Economy in Berlin', unpublished MA diss. (Humboldt-Universität Berlin).

Seidel, V. (2005), 'Best-practise-Analyze zum Quartiersmanagement mit dem Fokus "Integration von Migranten"', in Schader Foundation (eds).

Spaan, E., Hillmann, F. and van Naressen, T. (eds) (2005), *Asian Migrants and European Labour Markets* (London: Routledge).

Wilpert, C. (2003), 'From Workers to Entrepreneurs? Immigrant business in Germany', in Kloosterman, R. and Rath, J. (eds), *Venturing Abroad: global processes and national particularities of immigrant entrepreneurship in advanced economies* (Oxford: Berg).

Zukin, S. (1995), *The Cultures of Cities* (Oxford and Cambridge, MA: Blackwell).

Chapter 7

Social Innovation, Reciprocity and the Monetarization of Territory in Informal Settlements in Latin American Cities

Pedro Abramo

Introduction

During the past two decades, access to urban land for the poor in Latin American cities has been achieved primarily through the informal land market. In all major Latin American cities, the informal land market has been instrumental in the settlement of the neediest populations. And yet, that gateway has turned into a genuine social barrier for many urban dwellers in Latin American cities and access to land by the urban poor has become a key issue in the emerging urban agenda. Recent social indicators point to a major two-way movement of low-income urban settlements in Latin America. On the one hand, city outskirts are spreading outwards and 'extending' the metropolitan perimeter. On the other, there is a strong 'densification' of existing lower income settlements, pointing to a 'return' of the poor to the inner city. At both ends of this urban restructuring process, the informal land market acts as the key driver of land use change. Therefore, understanding the functioning of this informal market has become a central concern for urban policy makers and local administrators.

This chapter discusses some dimensions of the emergence, consolidation and transformation of what is considered one of the most important social innovations in Latin American cities: the rise of the informal city. In Latin America, the practice of collective land occupations – spontaneous or organized – and their transformation into a popular *habitat* can be considered an important social innovation of the poor in accessing urban land. The 'founding act' of land occupation must allow the emergence of other enabling social innovations for the production of basic goods, such as urban infrastructure and collective services, required for the consolidation of that space and its transformation into a popular territory. Self-production of housing and urbanization processes, initiated by a communitarian type of collective action transforms occupied land without urban infrastructure into an informal popular settlement that is effectively the territory (habitat) of social reproduction of the poor. This territory is the result of social innovations which accumulate and gain strength over time. These innovations are diffused and transformed into recurrent practices of the urban poor in the production of their daily urban life spaces. Contrary to the expectations of early institutionalist authors (Commons 1924), those practices do not transform the legal system but remain illegal or informal and, consequently, 'local authorities' are set up to mediate in conflicts and practices in informal settlements. Informal cities are, therefore, regulated by their own institutions, which often operate on the margins of the state. At the same time, the reproduction of social and economic life in those settlements depends on reciprocity and redistribution economies of a Polanyian type where the market is viewed as an auxiliary economy (Polanyi 1957).

Building upon this characterization of the institutional dynamics of informal settlements, this chapter discusses the emergence of a land market that is soon transformed into the primary coordinating mechanism of residential mobility, replacing both occupation processes and/or reciprocity economies as the primary regulatory means of residents' access and exit from these territories. However, the market that evolves within these informal settlements does not abide by the formal institutions that regulate market transactions. Every market transaction involves an exchange of property rights (Commons1924), but market transactions are also ruled by other rights such as commercial, urbanistic or environmental rights. In the case of the informal land market, land or housing transactions are carried out without any regard for property, urbanistic or even environmental rights. Thus, land markets in irregular settlements are of an informal nature; that is, the regulatory institutions of market transactions are of a different nature from those integrated in the legal system of the state, the formal market. But, since all markets require institutions to regulate contracts, conflicts or, in the case of housing, the continuity of the informal sequence chain as a coordinating mechanism, the informal market integrates those institutions that were built up socially during the process of territorial consolidation. In other words, the informal institutions established on the basis of reciprocity and distributive economics are now 'sucked up' by the informal land market which transforms them in its regulatory institutions.

This chapter examines two dimensions of the conversion of social innovations of the informal city into market valorization elements. The first dimension relates to the individualization of innovations in the form of location preferences and is developed in the following section, where I discuss some aspects of the construction of familial rationalities in the context of transforming a social innovation in its market and monetary expression.

A second dimension of this process of the 'sucking up' of innovations is analysed from the viewpoint of the 'locational capital' of the poor. In this case, a social innovation – the informal habitat – is converted into capital as a result of changes taking place in the intra-urban structure which valorize new accessibility factors. This dimension is taken up in the third section.

Location Preferences of Slum-dwellers

There is substantive empirical evidence for the increasing significance of the informal real estate market in determining access to land in slums. There is also considerable evidence that housing prices in the informal real estate market tend to be manifestly higher than prices in the formal market. On the basis of these patterns, the recent trajectory of urban poverty dynamics in large cities in Brazil can be examined through two critical questions. The first concerns the rationale of housing demand in slums: why do families choose to purchase housing in slums at such high prices? The second concerns the functioning of real estate markets and price formation in slums.

The literature on land and real estate markets defines a hypothesis of price competition as a determining factor in urban land market dynamics. This hypothesis implies that competition between formal and informal markets would render lower prices in the informal market because of more attractive alternatives in the formal one at equivalent prices; for instance, the substitution of residential proximity to the labour market for bigger houses on the periphery and the advantage of property titles.[1] However, recent research reveals that, in reality, competition between the formal and informal real estate markets does not take place.[2] The section below on the informal real estate market and investment strategies of slum dwellers offers some answers to this apparent paradox. In order to explain high(er) housing prices in slums, I examine the links between the labour market and the real estate market and maintain that the informal character of the first acts as a barrier to entry to the second's formal market.[3] Moreover, given the informality of the labour market and the relative rigidity of housing supply in the real estate market in slums, barriers to entry in the formal real estate market lead to the formation of a 'rationed market' (see also Mankiw and Romer 1992; Dixon and Rankin 1995)[4] which tends to respond to demand fluctuations with marked price variations.

Over the last thirty years, the characteristics of both labour markets and real estate markets have radically transformed the strategies of poor families trying to gain access to urban land and housing. Research on employment informality and

---

1   Choice dynamics among slum dwellers for residential options in the formal and the informal market with different characteristics and locations but similar prices were simulated in field work.

2   Preliminary results of field research about the real estate market (Abramo 1998; 1999a) are strongly conclusive as to the preferential choice of real estate property in the slums.

3   The locational terms of this relationship are but poorly explored, since the majority of the residential location models go along the lines of the Alonso-Fujita hypothesis of central location of the labour market. Likewise, the majority of labour market models are non-spatial. More recently, Zenou (1996) has tried to promote an articulation between the theories of labour market segmentation and the residential location models so as to attempt an explanation of the dynamics of urban social-spatial structuring.

4   The Neo-Keynesian concept of rationed market is used here.

real estate informality has revealed the extraordinary social heterogeneity among families drawing their incomes from means other than regular wages and/or living irregularly or illegally according to urban legislation. This social heterogeneity of the poor was revealed recently by a study on slums based on census-tract sections (Preteceille and Valladares 2000). Likewise, preliminary results from research on real estate markets in Rio de Janeiro's slums show a pattern of relative diversity among them as well as substantial internal diversity within them, a configuration that provides some basis for explaining the processes of real estate market segmentation within the slums.

On the other hand, fieldwork interviews among slum dwellers have also confirmed the significance of two critical factors in shaping the residential location preferences of the urban poor: proximity to eventual income sources and neighbourhood factors. However, interviews with second-generation residents provided interesting variations on residential choice. For the younger generations, the notion of proximity to income sources loses its territorial dimension *strictu sensu* in favour of a relationship network dimension; an eventual income prospect would be tied to the variety of opportunities provided by the network (religious, parental or personal relationships). In territorial terms, this network manifests itself in a more dispersed manner. Thus, an employment opportunity is not necessarily linked to physical proximity to a labour demand source such as a factory (Nova Brasilia in the 1940s and 1950s, and FernãoCarmim in the 1950s and 1960s), industrial or construction sites (Rocinha and Vidigal in the 1970s and 1980s) or household labour market (the 'ZonaSul' favelas in the wealthy southern neighbourhoods of Rio de Janeiro). A relationship network built around evangelical churches, for instance, opens up new sources of eventual income completely diffused in the city's territoriality. This feature of greater dependency of eventual work opportunities on social relationship networks is one of the new factors introduced in the decision-making process involved in the location preferences of slum dwellers. Nonetheless, this new social 'organic' element must not be overrated; data on income source location reveals that physical proximity factors are still quite relevant and a significant proportion of family income is obtained in the proximity area.

While territorial proximity is often essential in providing eventual income opportunities, the criterion for assessing the significance of proximity between place of residence and a potential employment source is of a different nature than the usual transport cost reduction factor.[5] A good example of this 'new' type of locational dependency is observed in the need for drug traffic activities to recruit

---

5   The transport cost reduction factor is defined by neoclassical models as the key explanatory variable to residential location choices by the poor. A two-fold transportation dynamic could be currently suggested to be verified. Its first aspect pertains to the movement undertaken by the very poor towards a 'superperiphery', very distant from the centre. The second aspect deals with the access by the poor to more central areas such as the slums. Being central is for these areas a determining factor on account of the weight of

their labour force from within the community where its base is located (Zaluar 1998; 1994). Trust and knowledge of the territory are, in this case, determining factors for establishing traffic networks in the slums. Both these traits emerge out of proximity relations and can be secured by recruiting labour from the locality where drug dealers operate. In this case, we are confronted with an eventual source of income for the slum dwellers which requires a proximity relation between place of residence and the development of such activity. Nevertheless, over the last thirty years, the relative importance of the proximity relationship has changed significantly.

Recent research on the location of income sources of slum dwellers in Rio de Janeiro (Municipal Work Office of Rio de Janeiro 1999) and São Paulo (Baltrusis 2000) has shown that a significant proportion of residents worked inside the slum itself. Thus, the slum can also be seen as a site where the concentration of informal services and retail activities generates a resource flux that constitutes an 'economic circuit' internal to the slum, which in turn feeds the local real estate market. This endogenous economic dynamic is established on the basis of such factors as trust and reciprocity. These factors mediate the interaction between social subjects and are critical components of the dynamics of local markets, which explains its endurance *vis-à-vis* the formal markets at lower prices. But this endogenous economic circuit acts to reinforce the significance of proximity factors in the choice of residential location. My research results are still preliminary but they suggest that empirical studies on the internal economy of slums may open up new horizons in the assessment of urban poverty trajectories in Brazilian metropolises, especially in terms of changes in the locational preferences of the poor and their relation to the location of their (eventual or effective) income sources.

The concept of proximity in slums may be effectively qualified using three different dimensions developed by 'proximity economics' theorists (Rallet 2000). The first definition of proximity is 'topographical': physical proximity. In this sense, the structuring principle of economic relations in the favelas is physical propinquity; that is, the slum's territoriality (its topography and location) provides for a kind of proximity that encourages economic and real estate activity.

A second definition of proximity is of a classificatory type: the slum and its dwellers acquire certain 'proximity' from the fact that they are socially grouped as residents. Given the fact that slums are linked by a particular trait of the land occupation process, that is, that there is no property title,[6] a form of proximity is established on a metaterritorial level. This proximity manifests itself in a territorialized way but still allows for a proximity connection between slums dispersed through the urban fabric. In other words, proximity in these slums is

---

transportation costs to familial overall income. Informal real estate markets thus reveal this process of slum densification.

6   Generally speaking, the land occupation process defines legal procedures particular to fundiary property. The slum is defined by its houses not having property certificates. For definition and discussion of this, see Saule (1999).

defined not by territorial contiguity but by a political and legal definition that establishes a 'new' urban territoriality: the slum.

The third type of proximity is 'organized proximity'. This form of proximity is established on the basis of institutionalized relations, informal or tacit networks and hierarchies, or even by sheer force and violence. The distinguishing trait of organized proximity is that it is a social construction involving agents and is maintained by a set of actions, norms, rules and procedures. Here, proximity is not geographical or classificatory but rather socially reproduced by means of a set of interacting relationships between individuals, families, groups and so on. Thus, organized proximity requires a kind of 'social sustenance'. Indeed, a multiplicity of 'organized proximities' account for the reproduction of economic and real estate relations.

Differentiating among these three types of proximity broadens the horizons for understanding economic phenomena in slums and makes it possible to articulate the dynamics of the real estate market and residential mobility in slums, with a set of other social practices that make up that specific territoriality.

Transforming Urban Territory and Locational Preferences

Residential location is generally considered one of the most important decisions in the familial universe of the poor, a perception that has been confirmed by empirical research. Residential location provides for differential access to sources of employment and income, service and retail centres, collective transportation, and public services associated with the positioning of the slum in the locational hierarchy of the city. Over the last thirty years, basic infrastructure developments (water supply, sewer systems, electricity supply) have contributed considerably to improved living conditions of slum dwellers. This trend is confirmed both by qualitative as well as by recently published disaggregated socio-demographic indicators for slum areas (Preteceille and Valladares 1999).

However, while social indicators express that there has been some progress in absolute terms in slums, the pattern of spatial distribution of public infrastructure and services has moved in the direction of strengthening social differences (Marques 1998). The benefits of public intervention in the local sphere are directly tied to the location of public investments that, in the last thirty years, have strongly favoured high-income areas of the city. Generally, these affluent areas concentrate natural positive externalities of the city, which are further advanced by substantial public investments. The resulting effect is an increase in intra-urban differences. The fact that some slums are in close proximity to these affluent areas makes it possible for them to absorb part of those externalities, a dynamic that also contributes to inter-slum differentiation.

Inter-slum differentiation reflects, to a large extent, the dynamics of change in the locational hierarchy of the formal city. This process may thus be seen as *passive* inter-slum differentiation, as opposed to *active* differentiation resulting from public sector urban initiatives (often highly selective) and/or self-improvement community actions. These public and/or community initiatives have an effect on the slum's physical structure resulting in localized improvements that lead to differentiation of the built environment of a city's slums. Distinguishing conceptually between these two patterns of inter-slum differentiation, passive and active, is essential for understanding the different dynamics of production of the built environment and their articulation with formal and informal territoriality.

Previous studies (Abramo 2001) have identified the dynamics of spatial structuring as a process of continuous differentiation of the built environment, marked by a spatial diffusion of spatial innovations and resulting in a counter-tendency towards uniformity. When inter-slum differentiation dynamics, both active and passive, are taken into account, the process of intra-urban structuring becomes more complex. This complexity calls for an analysis that takes into account patterns of continuity/change and appropriation/rejection relations between formal and informal urban areas. These relations determine dynamics of valorization/devaluation of the built environment and its externalities. In this way, the slums will be subject to valorization/devaluation as a result of changes in the formal built environment, but the latter would also be subject to requalification derived from changes in the built environment of the slum.

It can be said, then, that the built environment of the slum and its externalities are transformed over time on the basis of changes that are taking place in the slum itself. But, at the same time, change in the slums occurs as a reflection of changes in the formal environment of the neighbourhood. These transformations occur at different scales (street, district, administrative region, zones and so on) and their primary vector of change is the formal dynamic of transformation of the built urban environment. However, this vector of change, to the extent that it is viewed by the production and consumption agents of the formal structure as a 'negative externality' of the city, establishes a 'conflictive interaction' with the informal built environment.

Thus, the articulation between formal and informal processes of transformation in space and spatial differentiation (formal and informal) in the city is rather complex, projecting (on the basis of an spatial-temporal analysis) a genuine representation of a kaleidoscopic city.[7] In the first case, transformation dynamics in the slum are a manifestation of a dynamic of production of the urban built environment that results from informal relations (land occupation and self-improvement processes, real estate market activity) and/or public intervention on the part of NGOs. Nonetheless, the characteristics and existing hierarchy among slums also change on the basis of the character of their location within the urban structure. To a large extent, the urban structure of the area surrounding a slum and its position relative to other areas of the city are subject to constant transformation

---

7   Abramo (1998) presents a discussion on the configuration of the kaleidoscopic city taking into account only the formal processes of production of the built-on environment.

driven by the formal real estate market, public intervention and other agents intervening in urban materiality.[8]

Thus, the relative location of slums in the externalities map of a city, in terms of a hierarchy of attributes between slums (inter-slum preference relations), is transformed over time. A slum, while maintaining its attributes, may exhibit an improvement/worsening of its relative position due to territorial changes within the city. This reflective quality of slums transforms the slum dwellers' residence into a 'locational capital' that is valorized/devalued in the course of time – exactly as in the formal real estate market. Family strategies incorporate this factor in their inter-temporal budgetary calculations, tailing the evolution of the relative position of their 'locational capital' (their real estate property) within the intra-urban hierarchy and endorsing, in this way, the potential benefits/losses of eventual territorial shifts of the family residential unit (residential mobility) in the intra-urban structure. Therefore, a repositioning of this locational capital may stand for both upward mobility and a falling standard of living within a family.

An interesting observation is that a falling position in the urban location hierarchy does not necessarily mean a decline in a family's welfare, or *vice versa*. For instance, a family's decision to transfer its residence from a good location in the intra-urban hierarchy to the urban periphery may provide for monetary gain to finance a bigger house to accommodate all family members more satisfactorily. This family may also have anticipated that their new location will, in the near future, be in a superior position in the accessibility chart, due to future alterations of the city's road system. The informal real estate market is the mechanism through which locational factors of slum dwelling are transformed into locational capital, allowing the slum dwellers to enter a speculative game of wins and losses promoted by the transformations of intra-urban structure (see Abramo 2001).

This simple example illustrates the possibilities for slum dwelling families to increase their family welfare by exploring the alterations that take place in the city's territoriality, that is, in the urban built environment and its externalities. This general observation is confirmed when a person living in the Guaporé housing complex[9] declares that one of the most significant improvements that took place in his living environment within the last thirty years was the construction of a shopping/service centre in the surrounding formal neighbourhood:[10]

---

8   I believe a more comprehensive analysis of the urban structuring processes must take into account both the processes of formal/informal production of the built-on environment and the behavioural microdimension involved in the familial location decisions. In an ongoing work, I seek to develop this matter so as to redefine the general features of the kaleidoscopic city, incorporating the informal built-on environment production.

9   The Guaporé housing complex is located in the Penha district, in the suburbs of Rio de Janeiro city, and was erected in the 1960s to shelter those who were dislodged from the Catacumba slum. Nowadays, this housing complex is surrounded by three slums that are mostly inhabited by people related to those living in the Guaporé.

10   All quotes from the interviews were translated by the author.

Before … when I wanted to buy something I had to go too far away … in the last
few years the whole district has gotten much better and the Vila daPenha looks
almost like Copacabana, with that commerce, and the bars where we go to have
some fun on the weekends.

The sudden concentration of services, leisure and commercial activities (externality)
in the district contiguous to the housing development is thus appropriated by the
development as well as by the surrounding slums, as a valorizing attribute to their
locational capital, improving as a result the welfare of those families. In a similar
way, it is possible to identify areas whose decline has led to reduced locational
capital for poor families in informal housing adjacent to these areas.

As is known, cities display recurring changes in their urban functionality which
bear immediate relation to formal land uses. These alterations in the functionality
of urban areas affect the locational preferences of the poor. A good example is
provided by changes generated as a result of relocation of factory units within the
urban space. The preference of the poor for accessibility to employment sources
is well known and is often put forward to explain the emergence of slums (Abreu
1997).

The functional transformation of an area not only has an impact on the relative
position held by districts (and slums) in the city's accessibility chart; it also
alters its locational preferences. The slum, however, is an urban territoriality that
maintains strong primary relationship ties. The sentiment of identity associated
with belonging to a 'community' is generally seen by the slum dwellers as one of
the primary reasons directing a person's choice to stay in their original dwelling
place, as is shown by evidence drawn from the interviews. Familial synergy and
friendship bonds springing from many years of daily experiences within the slum
(the neighbourhood proximity effect) give rise to exchange relationship processes
based on the 'gift'/'counter-gift' mechanism that allows the maintenance of an
expanded 'familial solidarity economy' (Caillé 1994; Goudbout 1992). This
solidarity economy plays, within the universe of urban poverty, the very important
role of raising/educating children of families whose members mostly participate in
the (formal or informal) work market. Neighbours and relatives temporally assume
responsibilities that would normally involve either governmental responsibility
or monetary compensation, such as nurseries, caring centres for the aged and
disabled and so on. This gift is returned at a later time, giving rise to vast, complex
gift/counter-gift exchange networks that underlie everyday social relations within
the community.[11] Locational decisions also involve evaluation of accessibility to
such networks of 'solidarity economy', manifesting as they do in a predominantly

---

11   Recent studies on drug-trafficking within slums (Zaluar 1998; Leeds 1998) reveal
that these relations may also be identified within the criminal organizations that operate
in the slums – some of these relations may even be a prerequisite for the establishment of
these organizations.

territorialized form. Transference from the slum may therefore mean leaving such networks.

Residential mobility among the poor may, thus, be said to consist of very critical decision making within poor families' inter-temporal strategies, especially given the major 'social opacity' of the gift/counter-gift networks. Since they are not organized according to explicit norms (tacit relationship), each network will have a set of traits peculiar to the particular slum in which it is based. The territorialized aspect of these relations may therefore be said to constitute major instances of 'informational asymmetry' in the moment that the decision-making process on familial mobility is started. Families who decide to leave a slum (to sell their real estate property) know that, on leaving, their place within a specific solidarity network will not be transferred to the eventual buyer. Likewise, the family, on being transferred to another dwelling place, must face the uncertainty of encountering a different relationship network, a process that incorporates a 'learning cost' for the assimilation of the new tacit agreement.[12] On taking into account the 'territorialized solidarity economy' it can be argued that neighbourhood externality presents itself as endowed with huge importance for poorer families, and since it constitutes a positive externality, it will probably be highly considered in decision-making processes on residential mobility.[13]

Another interesting feature of the locational preferences of families refers to the possibility of recreating within the slum some everyday elements of a rural/ small-town character which are impossible to reproduce in other metropolitan locations. As stated by a resident of the Vidigal slum when asked about violence:

> Violence is a reality ... the police go up and down all the time, and it may happen that the boys [people who work for the drug traffickers] shoot at them, and then a gun fight starts ... but this is nothing for me. For me this community feels like living in the country again ... I have a garden where I plant some stuff, and I can chat with my neighbours all the time just like I did up there in the country ... I like it here very much ... there is the city but I still live like I did in the country.

Thus, due to the peculiar features of their territoriality, the slums make it possible to maintain diverse temporalities and 'lifestyles' that otherwise tend to be driven away from the city. It is interesting to observe how second-generation immigrants tend to absorb urban culture, forsaking their parents' traditions and habits. However, the slum also propitiates new living styles for the young, who maintain the primary communitarian feature of social relationships. The slum is both the space of daily life for the parents or grandparents who gather to talk by the little

---

12   This feature may explain research results in some slums where the largest residential mobility was found to be intra-slums (Abramo 1999b; Baltrusis 2000).

13   The value of solidarity as a positive externality will also be shown in housing prices in slums.

bars (local microcommerce), and the place where the night life of funk parties or rap group gatherings takes place, thus making for youth-groups' formation to which the proximity factor propitiated by the slum represents a strong cohesion element (see Zaluar 1998).

In this way, slum territoriality is turned into a true locational capital, be it in terms of the improvement of the slum's place within the accessibility hierarchy of the city as a whole, of neighbourhood externalities (accessibility to solidarity economies) or of 'urban welfare', making for the coexistence of different lifestyles within the slum territory. However, locational capital may be devalued by drug-trafficking activities and related violence. Paradoxically, the location choice carried out by the drug-trafficking feeds on the very same attributes behind familial decision-making: the contrast between well-placed territories within the accessibility hierarchy of the city and reduced internal accessibility (poor transportation features in terms of streets), or the gift/counter-gift economy of reciprocal exchange within which the drug trafficking plays its role through donations that fill the gaps left by scarce public support. Another important locational factor upon which the drug trafficking feeds is the 'territorial cohesion' element. A territorialized community with strong communal bonds that work to unite local population generally reacts as a whole to aggressions by external enemies (the police, rival drug cartels). These solidarity economy bonds constitute positive territorial externalities for the slum, but when appropriated by the drug-trafficking organizations the result is the devaluation of familial locational capital.

These negative/positive externalities, internal to the slum, add to the externalities of its surroundings in the real estate market devaluation process. Fieldwork carried out on the familial universe of slum dwellers suggests that a house constitutes the chief material investment for poor urban families. However, this house also represents a kind of property that has the feature of incorporating, or giving, individual access to urban externalities produced by collective action. During the last thirty years, these activities have constructed a very unequal pattern in the distribution of urban services and infrastructure in the city of Rio de Janeiro and the peripheral municipalities. Unequal spatial distribution is expressed in quantitative, as well as qualitative, terms through a concentration of equipment and services in certain districts, but also in terms of inequality, for the networks attending to poor urban areas are very different from those attending to richer areas. Besides, these networks, throughout the last thirty years, have evolved in a very aggressive way (Silva 2000; Marques 1998; Vetter and Massena 1981).

Within the universe of urban poverty, the traditional relation between residential location and location of the workplace (or probable income source) is still one of the key elements around which an explanation of the spatial distribution chart of the poor residences is to be constructed.[14] However, the spatial segmentation

---

14   Research carried out in several slums of Rio de Janeiro (Abramo 1998) showed that a high proportion (30 per cent) of heads of households were employed in districts contiguous to their residences.

found among service and public equipment networks introduces another variable in the locational decisions of families. In the words of a community leader from the Nova Brasilia slum, Rio de Janeiro:

> In the past the slum was surrounded by factories, and the people came to live closer to their workplace and to save transportation money, but now that the factories are all closed, we realize that being close to the commercial zones, and to the hospitals, is perhaps even more important than having the factory [close] by.

Thus accessibility to urban externalities, organized according to a very unequal distribution, becomes an important variable in the qualification of popular housing. Accessibility to public welfare institutions, as well as to natural externalities, is very important, for these constitute the free leisure that structures the daily universe of those families. The two different dimensions appear very clearly in the statement by an ex-inhabitant of the Catacumba slum who was transferred to the Quitungo complex, which is situated in a far-off periphery of the city:

> In the beginning, when we came here from the Catacumba, it was difficult to get used to these buildings with nothing around them. It was a little bit sad to see, because up there in the slum we had a beautiful view of the Lagoon, and the beach was also very close by. My children who were used to staying at the beach all the time had a hard time figuring out what they could do instead.

In this sense, residential location of poor families is an important element in their familial strategies, starting with transportation matters, but taking into account many other important elements constituting the pool of familial consumer goods. Another feature of the poorer families' universe of 'utilizable' urban externalities relates to their eventual relationship to much richer familial groups. This proximity between the slum and high- or medium-class families is normally regarded as a somewhat uncomfortable relationship, as is often seen in the mainstream literature on intra-urban structure (see Fujita 1989; Abramo 1997). The effects of neighbourhood externality, produced by identity and/or social homogeneity (of an ethnic, cultural, economic sort) of the richer families are constrained when a slum appears in the proximity of their dwelling places. Generally speaking, the 'estranging effects' manifest as prejudice on the part of the richer groups, so that a socio-territorial isolation of the slum is possible. However, the collected statements point to an interesting element produced by the proximity between social groups of different income levels, which allows us to analyse the common use of public space and of certain public equipment. Elements of the urban culture could be recalled (samba, football, cooking and so on), which approximate sociologically different groups. But it must be stressed that proximity between the slum and the workplace (be it work of a domestic nature, or linked to small-scale service delivery) allows a certain contact between those families dwelling in the slums and the richer families, which often includes the use by poorer children of a socio-economic space pertaining to richer children. The possibility of having access to higher educational levels, even when limited to childhood, is seen by most of the interviewees as a very positive element in the educational history of their children, being directly related to the locational factor of the dwelling place of the poor. This is clearly expressed by another former inhabitant of the Catacumba who was transferred to the Guaporé:

> I was very sad for my children when I came here, for they could always drop by the house where I worked [as a servant], which was just close to the slum. Over there they could learn the kind of good manners that I would never be able to teach them.

To sum up, for poor families, residential location is a key factor in enhancing or diminishing family welfare levels to the extent that it provides more or less accessibility to service delivering and public infrastructure networks. But it is also a factor that intervenes in the formation of slum dwellers' human capital.

Concluding Remarks

The production of informal cities and of the social reproduction spaces (habitat) of popular sectors constitutes the primary social innovation in Latin America's large urban centres in the second half of the twentieth century. In this chapter, I have examined some critical aspects related to the emergence, consolidation and transformation of these innovations, highlighting the risks posed by their appropriation by a market logic, notably in relation to the increasing monetarization of key community traits and interpersonal reciprocity dynamics within land markets. Notably, I have addressed two aspects of this process of market appropriation of social innovations in informal settlements. The first refers to the process of absorption by means of the conversion of the fundamentally collective character of social innovations to an individual value or dimension. The construction of the locational preferences of families is one way in which this individual valorization of externalities generated in the collective process of production of the informal habitat takes place. And it is precisely those locational preferences of the poor that are expressed in the informal land market, sold by those who hold them and transferred through the market.

The second aspect of this appropriation by the market of innovations of production of the popular habitat relates to the relation between that informal territory and the production of the formal city. The location of that socio-spatial innovation and its characteristics in terms of infrastructures and services should not be considered in absolute terms, but rather in relational ones. The territorial location of the informal settlement and its valuation within the hierarchy of accessibilities changes with the process of urban transformation and carries within it the virtual capacity to improve (or degrade) its relative position in the vectors and intra-urban characteristics of the city's growth. From this perspective, the location of a social innovation over time can be transformed into locational capital for its residents. A collective process of production of an informal habitat, classed here as social innovation, is transformed into a virtual capital, a 'locational capital' that incorporates a monetary dimension in the informal land market. The relational dimension is key to the transformation of the material output of collective action into an individual value of a relational character, since a piece of land, a plot (a piece of the territory of innovation) only has monetary value relative to its position in the urban accessibility hierarchy.

The two dynamics analysed show the same dimension of market individualization of material (self-built housing and self-urbanized spaces) and/or social (relations and reciprocity and vicinity economies) outcomes of collective actions whose starting point is a social innovation. Here we are confronted with a paradox: a social innovation that is developed during the process of a collective break with the market is then appropriated and turned into a commodity to be traded in the informal market. This paradox makes it possible to recognize that the trajectory of a social innovation may at any point be altered and sucked up by a logic of individualization of interests and its mercantile manifestation as a market exchange. This contingency warns against absolutist interpretations of social innovation and recommends a thoughtful evaluation of normative applications in public policies and strategies.

References

Abramo, P. (1997), *Le marché et l'ordre urbain: du chãos à la thèorie de la localisation résidentielle* (Paris: L'Harmattan).

Abramo, P. (1998), 'Impacto do Programa Favela-Bairro no mercado imobiliário de favelas da cidade do Rio de Janeiro', unpublished research report (Rio de Janeiro: IPPUR-UFRJ).

Abramo, P. (1999a), 'A dinâmica do mercado imobiliário e a mobilidade residencial nas favelas do Rio de Janeiro: resultados preliminares', unpublished research report (Rio de Janeiro: IPPUR-UFRJ).

Abramo, P. (1999b), 'Formas de funcionamento do mercado imobiliário em favelas', unpublished research report (Brasilia: CNPq).

Abramo, P. (2001), *La Ville Caleidoscopique* (Paris: Harmattan).

Abreu, M. (1997), *Evolução urbana do Rio de Janeiro* (Rio de Janeiro: Ed. Zahar).

Baltrusis, N. (2000), *A dinâmica do mercado imobiliário informal na Região Metropolitana de São Paulo: um estudo de caso nas favelas de Paraisópolis e Nova Conquista* (Campinas: PUC).

Caillé, A. (ed.) (1994), *Pour une autre économie* (Paris: La Decouverte).

Commons, J.R. (1924), *Legal Foundations of Capitalism* (Basingstoke: Macmillan).

Dixon, H. and Rankin, N. (1995), *The New Macroeconomics: imperfect markets and policy effectiveness* (Cambridge: Cambridge University Press).

Fujita, M. (1989), *Urban Economic Theory: land use and city size* (Cambridge: Cambridge University Press).

Gilly, J.-P. and Torre, A. (eds) (2000), *Dynamiques de proximité* (Paris: Harmattan).

Godelier, M. (2007), *Au fondament des sociétés humaines* (Paris: Albin Michel).

Goudbout, J. (1992), *L'esprit du don* (Paris: La Decouverte).

Henriques, R. (ed.) (1999), *Desigualdade e Pobreza no Brasil* (Rio de Janeiro: IPEA).

Leeds, E. (1998),'Cocaína e poderes paralelos na periferia urbana brasileira', in Zaluar and Alvito (eds).

Machado, L. (ed.) (1981), *Solo urbano: tópicos sobre o uso da terra* (Rio de Janeiro: Zahar).

Mankiw, G. and Romer, D. (1992), *New Keynesian Economics: imperfect competition and sticky prices* (Cambridge, MA: MIT Press).

Marques, E. (1998), 'Infra-estrutura urbana e produção do espaço metropolitano no Rio de Janeiro', *Cadernos IPPUR* ano XII:2, 57–72.

Maurin, E. (2004), *Le guetto français: enquête sur le separatisme social* (Paris: Seuil).

Municipal Work Office of Rio de Janeiro (1999), *Novas favelas na cidade do Rio de Janeiro: uma primeira aproximação a partir de levantamento aerofotogramétrico* (Rio de Janeiro: IPLAN-RIO).

Polanyi, K. (1957), 'The Economy as Instituted Process', in Polanyi, Arensberg and Pearson (eds).

Polanyi, K., Arensberg, H. and Pearson, H. (eds) (1957), *Trade and Market in the Early Empires: economies in history and theory* (New York: Free Press).

Preteceille, E. and Valladares, L. (1999), 'A desigualdade entre os pobres – favela, favelas', in Henriques, R. (ed.).

Preteceille, E. and Valladares, L. (2000),'Favela, favelas: unidade ou diversidade da favela carioca', in Ribeiro, L. (ed.).

Rallet, A. (2000), 'De la globalisation à la proximité géographique', in Gilly and Torre (eds).

Ribeiro, L. (ed.) (2000), *O futuro das metrópoles* (Rio de Janeiro: Ed. Revan).

Saule Jr, N. (ed.) (1999), *Direito à cidade* (São Paulo: Max Limonad).

Silva, R. (2000), 'A conectividade das redes de infra-estrutura e o espaço urbano de São Paulo', in Ribeiro (ed.).

Vetter, D. and Massena, R. (1981), 'Quem se apropria dos benefícios líquidos dos investimentos do Estado em infra-estrutura?', in Machado (ed.).

Zaluar, A. (1994), *Cidadãos não vão ao paraíso* (Rio de Janeiro: Escuta).

Zaluar, A. (1998), 'Crime, medo e política', in Zaluar and Alvito (eds).

Zaluar, A. and Alvito, M. (eds) (1998), *Um século de favela* (Rio de Janeiro: Fundação Getulio Vargas).

Zenou, Y. (1996), 'Marché du travail et économie urbaine: un essai d'intégration', *Revue Economique* 47:2, 263–88.

Chapter 8

Social Innovation and Governance of Scale in Austria

Andreas Novy, Elisabeth Hammer and Bernhard Leubolt

Social Innovation and Governance of Scale

This chapter studies the relationship of social innovation and space in Austria from a historical perspective, giving special emphasis to two emblematic moments in periods of crises of hegemony. It mobilizes empirical research to reflect on crucial questions concerning social innovation and re-conceptualizes social change dynamics and scalar arrangements.

Transformations in socioeconomic development often go hand in hand with changes in time and space, as the construction of new institutions and regulations coincides with the production of space (Harvey 1996, 208ff.). Therefore, we analyse social innovation and the governance of scale in Austria in a historical perspective so as to be able to grasp the processes of their institutionalization and structuring.

The Dialectics of Content and Process in Social Innovation

Social innovations as creative bottom-up initiatives are important aspects of social transformation, as they are often the first steps towards its institutionalization. As multidimensional processes they emphasize the necessity of overcoming a solely technology-focused approach towards innovation (Moulaert and Nussbaumer 2005). Although this is largely forgotten today, Schumpeter himself stressed social innovation as a necessary condition for the efficacy of technological innovation (1932), an insight that has been used for systematizing types of social innovation (De Muro et al. 2008). According to Moulaert et al. (2005), social innovation is comprised of dimensions based on procedure as well as on content. The procedural dimension primarily focuses on institutional aspects in the innovation process. New planning procedures, new forms of governance, empowerment and participation form the cornerstones of this dimension. The other dimension stresses the outcomes of innovation processes. Social innovation as a process of organizational change has to achieve objectives in enlarging entitlements, satisfying basic needs and increasing the quality of life as well. If these two dimensions are pursued independently, this works to the detriment of integrated development (Sen 1999; Novy 2002). Our approach explicitly aims at integrating the procedural and the outcome-oriented dimension of social innovation: social innovations are those forms of democratic governance which enlarge the entitlements of all.

This approach has been enlarged in more recent work towards a multi-dimensional approach, which emphasizes the context-specific and path-dependent strategies taking into account the historical trajectory and the territorial and institutional setting (De Muro et al. 2008). *Process innovations* cover all attempts to foster empowerment, participation and socioeconomic democracy, and have the potential for lasting effects in terms of institutionalization. But they are prone to cooptation by hegemonic or dominant institutions, not least because of the way they are overruled by hegemonic discourse (Moulaert et al. 2007). *Content innovations* consist of creative strategies to enlarge entitlements and satisfy basic needs. In experimental and bottom-up initiatives this has been achieved in diverse and decentralized ways.

Social Innovation and the Governance of Scale

Process- and content-oriented social innovations have always been sensitive to scale. They take place as experiments in a specific location during a specific time. But if they are successful, they are amplified and generalized and become institutionalized. During the twentieth century, institutionalization was national. Therefore, analysis has to embed initiatives in overall societal, economic and political structures as well as respective power relations (Moulaert and Cabaret 2006). Different historical phases of socially innovative development tend to reflect aspects of somewhat broader societal and political shifts of capitalist development. This chapter continues reflections pursued in previous studies on the specificities of social innovation in Vienna (Becker and Novy 1999; Novy et al. 2001; Novy and Hammer 2007).

Welfare via Top-down Policies

The Habsburg Empire was a hybrid, containing elements of the liberal Western and the state-centred Eastern model. While Western Europe developed an urban citizenry early, and serfdom was abolished before the nineteenth century (Anderson 1978), cities and civil society have been suppressed in Eastern Europe, where domination of peasants continued until the twentieth century (Anderson 1980). Furthermore, counter-reformation restored absolutism and political Catholicism (Vocelka 2002, 115). The failed revolution of 1848 and the neo-absolutist conservatism caused difficulties for the liberals already in the nineteenth century. They were marginalized by the Viennese Christian Social Party which had a petty bourgeois, anti-Semitic and estate-based ideology (Pelinka and Rosenberger 2000, 20; Vocelka 2002, 220). After the First World War, 'German Austria', the territory that remained after secession of non-German nationalities and the foundation of diverse successor states, faced severe difficulties, particularly in the economic field.

Annexation to Germany was therefore a widespread desire which was forbidden by the victorious *Entente* (Hobsbawm 1990, 92; Berger 2007). Thus, the Austrian nation-state enjoyed little acceptance among its population.

Social Innovations Emerging from Civil Society

Austria was a late-comer to industrialization, and civil society was always weak and grew in strong symbiosis with the state. As entrepreneurs who were independent from the state were rare, liberals recruited their followers more from the educated classes than from national entrepreneurs. Free-market ideology was never very strong, and after the economic crisis of 1873, it lost even more of its appeal. Two other ideological camps occupied the wasteland of hegemony: the catholic-conservative, the 'blacks', and social democracy, the 'reds' (Pelinka and Rosenberger 2000, 20ff.). While the liberals were elitist, much in tune with the cultural elite of Vienna's *fin de siècle* of Freud, Schönberg and Klimt (Schorske 1982), the blacks and reds were mass movements with a proper political culture, the one based on Catholicism and the other on Enlightenment and Marxism, but always moulded and moderated by mass culture. The 'blacks' had a communitarian outlook, focusing on local roots and political Catholicism, whereas the 'reds' took a more collective and internationalist stance. Post World War I, a coalition of these two antagonistic blocs formed for the first time, though with clear class cleavages. They agreed on a republican constitution that contained the germ of a Welfare State and realized conservative ideas of 'federalization' to limit central power.

The Local State Project of Red Vienna in the 1920s

In Vienna, the Austro-Marxist social democratic party took over the city government. In 1922, the large state of Lower Austria was split into two parts: the city of Vienna, which became a proper state, and the surrounding countryside (Stimmer 2007, 13). As the national government was conservative, all intellectual and political efforts of the party and the labour movement concentrated on the city, a financially well funded 'red island' in the 'black sea' of the Austrian hinterland (Öhlinger 1993, 13). *Red Vienna*, the social democratic government of the city-state of Vienna, influenced the thinking of Karl Polanyi on embedding market societies. It was a dissuasive example for Friedrich August von Hayek and Ludwig von Mises, founder of the Austrian school of economics (Peck 2008), even though it was a reformist project of redistribution which accepted the capitalist laws of accumulation, because social democracy respected private ownership of the means of production. However, it lacked, a consistent economic strategy, the positive effects on local industry being mainly the result of their own social commitment. The federalist Austrian constitution led to greater room for manoeuvre, which was used to introduce innovative and world-wide praised strategies in such diverse fields as housing, school reform and social policy. *Red Vienna* was based on a strong link between the party, civil society and the city administration (Maderthaner

1993). Social democracy implemented democratic reforms in the internal structure of the municipal bureaucracy, the school and culture (Achs 1993; Böck 1993; Öhlinger 1993, 12). But it continued to sympathize with a model of development via trusteeship, modernization from above but in favour of the masses. This became especially clear with respect to housing, where self-organization was replaced by large-scale public housing (Pirhofer 1993). The construction of more than 60,000 public apartments resulted in socioeconomic homogenization within the city and in a more equal spatial distribution of wealth (Jäger 2003). Social democrats also incorporated and co-opted certain innovative movements, such as a powerful settlers' and cooperative movement, which represented a broader effort of social organization beyond state bureaucracy and the market (Novy 1993, 18). Cooperative housing emerged out of necessity (1918–1921), resulting in the dynamic creation of cooperative settlements (1922), the municipalization of the settlers' movement (1923–1930) and an emergency programme organized from above during the depression (1930–1934). Settlers contributed to 55 per cent of new public housing in 1921, 14 per cent in 1924 and only 4 per cent in 1925 (Novy 1993, 98). Step by step the settlers' movement lost its dynamics and the unity between architectural and socio-cultural innovation was lost (Novy 1993, 24).

The Austromarxists rejected violence and wanted to implement socialism via consciousness raising (Pirker and Stockhammer 2006). For them, internationalist socialism was the utopian horizon, concrete social policies that accepted liberal democratic rules and the national power container, the praxis. Their enemies were less noble. To undermine the redistributive power of the local state, the conservative national government started a deliberate strategy of centralization against *Red Vienna*. Its core was the so called 'fiscal march against Vienna' (Melinz 1999, 19), a strategy still enshrined within the constitution. Competences thus shifted from local and regional scales to the national. But in 1933 the governing Christian-Social party staged a creeping auto-coup, abrogating the constitution and leading to a short civil war in 1934 that ended liberal democracy and introduced a dictatorship inspired by Italian fascism and based on Catholic corporatism (Becker, Novy and Redak 1999, 4). Social democracy was declared illegal. In 1938, Austrofascism was displaced by the Nazi occupation and an even more severe form of dictatorship emerged. Vienna became the experimental field of Aryanization linked with rationalization in petty trade and production (Aly and Heim 1993). Two hundred thousand Viennese citizens were either deported to concentration camps and killed or had to flee the country (Faßmann 1995, 14).

Nation-state-led Social Reformism from the 1950s to the 1970s

The fertile grounds on which fascism, racism and authoritarianism flourished were not cleaned away after the Second World War (Aly 2005). Elements of the Nazi welfare regime were even used for the construction of the post-Nazi Welfare State (Becker 2000, 96). The Left and the Right opted for a consensual civil society model of social partnership in the post-war period, which accommodated the class struggle in democratic institutions. Elites opted for a one-nation strategy of the Keynesian Welfare State, 'in which the support of the entire population is mobilized through material concessions and symbolic rewards' (Jessop 1990, 211). State-led social reformism organized a successful consensus-oriented corporatist growth coalition. The two main parties were state and mass parties at the same time (Agnoli and Brückner 1967; Poulantzas 2002, 262), controlling civil society via a mixture of benevolence and cooptation.

Austria became one of the countries with the strongest corporatist arrangements worldwide. The 'social partners' – the national associations of capital and labour – became the main – in economic matters the sole – representatives of a corporatist mode of governance with liberal, catholic and nationalist, authoritarian and democratic traces. Social partnership was a form of societal planning and coordination based on the accommodation of antagonistic class interests through negotiation between the two opposed, but now cooperating, blocs. Although ideological dispute receded with increased wealth, it structured not only the state but also civil society, as every private organization had to be linked to one of the blocs (or to their respective associations and think-tanks) if it wanted to receive state resources. Social democracy and trade unions successfully aimed at containing or coopting all progressive initiatives that sprang up in civil society. This corporatist framework of social partnership integrated the reformist wing of the workers' movement into a coherent power structure, but marginalized more radical movements on the left (Becker, Novy and Redak 1999). The borders between government, parties and civil society were unclear and civil society was never able to develop as an autonomous sphere (Novy and Hammer 2007, 212). Social innovations as bottom-up initiatives, however, were limited, co-opted or repressed: 'Vienna is a literally lethal atmosphere for critical minds' (Novy 1993, 194). The collaboration of capital and the reformist wing of the workers' movement represented a hegemonic bloc without any significant opposition. It remained a stable growth coalition, even when the social democrats won an absolute majority in 1970.

This tradition of social reformism (Friedmann 1987, 87ff.) has deep roots in Central Europe with Neurath und Mannheim as key advocates of a rational and democratic planning (Johnston 1983). Social reformism emerged as a reaction against irrational policies as well as against totalitarianism. It can in itself be called one of the most important social innovations of the twentieth century. Policies were conceptualized 'from above' and put into practice by academic experts attempting to implement them; this often led to a paternalist attitude of trusteeship of the already developed over those awaiting development (Cowen and Shenton 1996, 4). Its uncontested main objective consisted in reconciling economic growth with social and regional equity. Based on a positivistic understanding of science and politics, development became a planned process in order to create a better society.

To sum up, the decades after the Second World War were the golden age of capitalism based on a national mode of development. The Keynesian national Welfare State was the result of a broad variety of socially creative strategies pursued

Table 8.1   Two ideological camps and corporatist social partnership in Austria

	LEFT 'the reds'	RIGHT 'the blacks'	SYNTHESIS Social partnership
State (executive)	Ministry of Social Affairs	Ministry of Trade and Commerce	Government
Party system	Social democracy	Popular Party	Legislative
Public Chambers	Chamber of Labour	Chamber of Business and Trade	Expert committees giving technical advice
Private social partners	Trade unions	Federation of Austrian Industry	Various institutions of social partnership
Private organizations	Various	Various	Church-based civil society, workers' movement-based civil society

Source: authors' elaboration.

at different places and scales, moulded by traditional and structural elements. Parliamentary democracy and the Welfare State were finally institutionalized at the national level, although experimentation took place at other scales as well. Fordist social reformism in Austria, organized by the national state and centralized corporatist institutions, modified but only marginally overcame the deeply rooted structures of an authoritarian, paternalist and patriarchal political culture created by counter-reformism, Josephinism and a strong state bureaucracy (Johnston 1983, 15ff.). Corporatism was an efficient and successful form of governance for fostering economic growth, but hindered social innovation by its top-down decision making and rigidities.

Socially Creative Strategies from Below

As the 1968 protest movement was relatively weak in Austria, reforms were once again implemented in general top-down fashion by the national government. While nationally oriented full employment politics were abandoned in many European countries, social democratic Chancellor Bruno Kreisky (1970–1983), recalling the Great Depression, openly admitted preferring an increase in government debt to an increase in unemployment. In 1979, an attempt to resist recessionary monetarist policies failed for the first time, when the Austrian Central Bank kept interest rates low and lost a third of its reserves (Unger 2006, 74). After 1979, the social democratic government applied a variety of strategies, which together were later termed Austro-Keynesianism. The combination of a fixed exchange rate for the German Mark and Keynesian deficit spending was a key element of

Austro-Keynesianism; later this was supplemented by low interest rates, the absorption of labour in nationalized industries and a policy of wage restraint. Trade unions moderated their demands in favour of a consensual strategy of state-centred corporatism (Novy, Lengauer and Coimbra de Souza 2007, 14ff.).[1]

The 1970s was the decade of legal and institutional reforms to broaden socioeconomic citizenship (Berger 2007, 326ff.). At the same time, and despite the top-down nature of the national government's agenda, the crisis of the post-war accumulation regime after 1979 was a conjunctural moment when alternative development became popular at the margins of the political system.

Social Experiments in all Fields of Society: A National Strategy of Social Innovation in the Early 1980s

In the early 1980s, socially creative strategies were promoted as an answer to the crisis of Fordism, which shook nation-state-centred regulation and its reliance on top-down engineering. Unemployment was combated not only by traditional Keynesian demand management, but also by new programmes for an incipient social economy and self-reliant development. Walther Stöhr, a leading scholar of urban and regional development in Austria, was one of the founders of 'development from below' or 'endogenous development'. This concept of self-reliant development was much in line with world-wide dynamics and elaborated as an antithesis to a top-down approach to development (Stöhr and Tödtling 1978; Stöhr and Taylor 1981; Friedmann 1992; Moulaert 2000, 62). It combined the aspirations of local political movements and their protest against exogenous, large-scale development projects and their struggle for a more self-determined development. Furthermore, it aimed at a transition from planning as state-led social reform to a social learning and social mobilization paradigm in three respects (Friedmann 1987). First, these aspirations can be qualified, using a broader term, as the struggle for *democracy*, *participation* and *empowerment* (Gerhardter 2000; Rohrmoser 2000, 234). Second, the well-being of a population was no longer defined as only material, but also in terms of other factors (nature, culture …). Stöhr thereby reflected upon both the discussion of *basic needs* and the *diversity* of social groups. The anti-economistic attitude, clearly aimed against neoclassical theory, is sometimes reminiscent of a romantic anti-capitalism (Stöhr and Tödtling 1978). Third, the rise of a *third sector* occurred, as regional development policies in Austria started to pay attention for the first time to the non-market-based sector within the capitalist economy, be it

---

1   Since this time, the rate of structural unemployment has continued lower in Austria than in the rest of Europe: between 1970 and 1999 while the average rate of unemployment in Europe was 6.4 per cent, in Austria it was only 3.3 per cent (Unger 2006, 68). In terms of inequality in income and consumption, Austria's good position has been confirmed in the UNDP reports on human development: whereas the poorest 10 per cent of the population in liberal governance regimes, such as Great Britain, earn 2.1 per cent of total income, Austria's poorest 10 per cent earn 3.3 per cent (UNDP 2006, 335).

self-managed or cooperative. Although resources were provided mainly by the central state, programmes were increasingly executed by actors of civil society at regional and local scales (Gerhardter 2000, 81). This endogenous development approach achieved an integration of both dimensions of social innovation in a convincing way: alongside the economic objective, that is, to secure employment in regionally disadvantaged areas, endogenous development programmes were intended to stimulate self-help and self-management. Such attempts at implementing endogenous development were complemented by approaches of *community work* and support by consultancy, which had already gained ground in the course of the 1968 movement. The demand for equality was accompanied by a visionary perspective of expanding freedom of agency for every citizen.

In the 1980s, Austria faced a profound process of economic restructuring, a shift that was dealt with using an open, dialectical approach to social policy. Neither the social nor the economic sphere was seen as a system with its own logic. On the contrary, the interdisciplinary definition considered society as a unity, but one structured by internal contradictions. Such a view foreclosed an approach that would establish a clear and logical relationship between means and ends. Diversity and plurality were seen as the means of searching for socio-cultural and organizational innovations in order to handle the new situation of increasing unemployment and slow economic growth. Policy makers understood the social economy in a broad sense. Initiatives were founded from below and supported by an emerging political and administrative elite, resulting in forms of social innovation oriented towards deprived groups traditionally neglected by power holders. It was an integrated and interdisciplinary approach to development.

Governance arrangements changed too. The national state encouraged and supported an emerging alternative project culture which aimed at reducing unemployment at the regional and local scales. The social democrat Alfred Dallinger, Minister of Social Affairs from 1980 to 1989, was crucial in this process. Social innovation remained, to a great extent, dependent on the strategies and power of a broad range of reform-oriented individuals. Nonetheless, Dallinger took important legislative steps to implement structural incentives for social innovation. An amendment was enacted in 1983 in order to foster the development of small-scale socially innovative experiments in the field of social economy and similar legislative initiatives followed. Dallinger, although deeply rooted in the tradition of social partnership, facilitated access to state resources for hitherto marginalized groups and small-scale initiatives. He tried to reform corporatist governance by broadening it. New actors in civil society, from dedicated individuals to grassroots initiatives and self-managed cooperatives, began to cooperate with the Ministry. The ecological, feminist and peace movements, together with citizens' movements and urban action groups, had emerged at the end of the 1970s. This went hand in hand with an upgrading of the local scale. Additionally, these groups put forward not only a different perspective of development, but also a non-corporatist way of conceptualizing politics and the state. They formed a new and alternative segment of civil society, hitherto unrepresented in the power networks. Dallinger allowed these actors to create and enlarge state spaces as public spaces of their own, softening the necessity of strict adherence to the 'reds' or the 'blacks'. In doing so, he helped to create a cultural and political milieu that permitted the emergence not only of a broad variety of NGOs, but also of an alternative party on the left, the Greens (Schandl and Schattauer 1996).

Containing Experimentation (1986–1994)

The conservatives returned to government in 1986, forming once more a great coalition of reds and blacks. In line with the post-war tradition, the Ministry of Social Affairs remained social-democratic, while the Ministry of Trade and Commerce became conservative. The new government substituted social liberalism for national Keynesian policies of demand management. While the conservatives dominated privatization and liberalization in economic policies, social democracy defended the breadth of the Welfare State while changing its organizational form. This strengthened the corporatist form of governance. Social-democrats were forced by conservatives and opinion makers to limit social experimentation especially in the field of alternative life-styles and policies. Fiefdoms of 'red' and 'black' influence were re-established. Small social initiatives could no longer stay independent, but were subsumed under the patronage of institutionalized non-profit organizations belonging either to the 'red' or to the 'black' power bloc. To avoid repartisanship, they had to professionalize and internationalize.

A number of initiatives in the field of social economy had grown in importance. In the field of culture, tourism and social work, this trend brought about the development of new job profiles and appropriate training facilities. In other cases, formerly subsidized social economy initiatives transformed themselves into for-profit organizations, as in the case of the *Österreichische Studien- und Beratungsgesellschaft* (ÖSB Consulting) and the *Österreichische Arbeitsgemeinschaft für Regionalentwicklung* (*ÖAR Regionalberatung* [ÖAR Regional Consulting]). As a consequence, the onus of social innovation shifted to those professionally employed within already existing organizations. Socially innovative ideas, thus, were no longer developed from below together with help from outside, but conceptualized predominantly by an emerging elite of professionals in third sector organizations and consultancy firms. A proper research institute for not-for-profit organizations was created at the Vienna University of Economics and Business Administration. Nevertheless, some institutes with an innovative, public and transdisciplinary approach were able to flourish, for example the *Zentrum für Soziale Innovation* [Centre for Social Innovation] (ZSI) and *Forschungs- und Beratungsstelle Arbeitswelt* [Working Life Research Centre] (FORBA).

Socially creative initiatives in the field of social economy were no longer directly financed by the Ministry of Social Affairs, but by the Public Employment Service (*Arbeitsmarktservice* – AMS), which was founded in 1994. AMS is self-managed by the social partners, but mainly funded by the Ministry of Social Affairs. It has a decentralized structure giving more voice to the social partners and regional governments. The social economy was institutionalized and streamlined through standardized proceedings. Social innovation, thus, was channelled into predefined directions via a reassessment of existing instruments. Entrepreneurship – innovation as short-term market success – was seen as decisive for overcoming the crisis of Fordism. Initiatives were streamlined, limiting the only recently created space for social experiments. Initiatives had to prove their conformity with the uniform standards of overall employment policies. In particular, social innovation in the field of culture, which was considered subversive, was only subsidized on rigid terms aiming at a de-politicization of these initiatives.

Social-Liberal Mainstreaming of Socially Innovative Experiments in Austria (1995–1999)

In 1995, Austria joined the European Union which implied a crucial transformation of the corporatist form of social partnership. Legislation and procedures were adjusted to the social-liberal mainstream at the EU level. A discursive and organizational order based on market, enterprise, commodification and competition became hegemonic (Novy 2002). This represents an emerging 'hegemonic consensus' (García and Claver 2003), which Oberhuber (2005) calls 'mainstreaming' in his discourse-analytical study of the drafting of the European Constitution. A 'stream' of communications is inconspicuously but steadily narrowed down, extremes on both sides are discarded, divergent questions and issues are marginalized, deviant positions ignored or ostracized; the stock of taken-for-granted assumptions that must not be questioned is thus accumulated and a dominant discourse (a 'mainstream') established (Oberhuber 2005, 177; Leubolt, Novy and Beinstein 2007).

Contrary to liberal argumentation, markets are not natural but artificial institutions, made by women and men (Polanyi 1978). The integration of the European market has been a political undertaking to free national economies from national regulations (Huffschmid 2007). In the 1990s, the neoliberal market ideology received a further impetus, going beyond 'freeing' markets from corporatist and democratic interventions by creating new markets in formerly non-commodified fields (Novy 2002, 124). In order to create markets, policy makers have to transform organizations into firms and convert all social relations into market relations. This process of the commodification of social policies and activities [*Ökonomisierung des Sozialen* (Bröckling, Krasmann and Lemke 2000)] consisted of three strategies. First, as regards profit and not-for-profit organizations and grassroots initiatives, these organizations had to reinvent themselves in a business-like structure if they wanted to be eligible for subsidies. These isomorphic dynamics were key factors in the replacement of the organizational diversity of the 1980s by firms or nongovernmental organizations working within the logic of New Public Management (Pelizzari 2001). Second, the quality of various public services and goods was transformed into clearly delimited commodities. Performance indicators and evaluation instruments have been crucial ingredients in transforming the quality of various activities into quantities, thereby making activities comparable. Third, comparability permits relating these activities to one another, thereby generalizing competition to the detriment of cooperation.

As regards social innovation, this new EU-imposed form of governance was ambivalent. One can observe redefinitions in both of the two dimensions of social innovations. The procedural dimension was emphasized at the same time as institutionalized organizations and firms were favoured, to the detriment of new experiments. EU programmes are based on a positivist understanding of politics. A logical framework prescribes the optimal procedure *ex ante*. Contextual specifics cannot be dealt with and the participation of target groups (the 'stakeholders') must not have any implications for the rules of the game, which are imposed from above. Nevertheless, EU membership was enthusiastically welcomed by the social sector and the current initiatives of the social economy. Non-profit organizations and the alternative milieu correctly perceived that EU rules of the game challenged the rather closed national system of corporatism. Market regulation has some important advantages over corporatist regulation: first, it permits the participation of an enlarged number of actors, thereby reducing the traditional dominance of the 'red' and 'black' milieus. Second, in line with positivism, it puts competition on an objective base, thereby depoliticizing the allocation of resources.

While these changes were being applauded by those involved in social initiatives, power holders started to redefine the objectives of social innovations. In a manner similar to Fordist social reformism, logical positivism has a proclivity towards top-down forms of governance, if not controlled by democratic governance. In the 1990s, the relation of economics to social objectives was turned upside-down (Gerhardter 2000, 83). The central strategy of the decision makers (the social partners or the Ministry) was to advance the commodification of social initiatives. The supervision of professional project management and the development of efficient funding applications became the main preoccupations of the Public Employment Service (AMS). Non-economic initiatives were only funded if they could prove their economic utility. Open processes and experiments could no longer pass the rigorous check of economistic *ex-ante* appraisals. Regional consultancy became widespread, as the regions themselves were merchandized, and consultancy was reduced to business or technological support without broader societal aspirations (Gerhardter 2000, 85).

Authoritarian Embedding of Social Liberalism (2000–2006)

In 2000, a right-wing coalition came to power in Austria and broke with social partnership. In the administrative structure of the Ministry, this was expressed by the partial annexation of the hitherto existing Ministry of Labour and Social Affairs to the Ministry of Economics and Labour. The traditional, consensual division between 'red' and 'black' was abandoned. However, the weakening of trade unions did not lead to the prospering of a pluralist civil society, but to a further reduction in the organizational diversity of actors. Self-managed firms, the voluntary sector and cooperative-like enterprises were on the retreat. Furthermore, the diversity of goods and services delivered was reduced, be it community schools, child-care facilities or subsidies to Third World groups. Cooperation, common in this field until then, was reduced to price competition or competition for subsidies. In all these aspects, the market has produced fragmentation, not pluralism.

With regard to the form of governance, however, the government adopted traces of patrimonialism, a form of state organization, which does not allow a clear distinction between the state as an organization and as the property of the rulers (Weber 1980). Paternalist and patriarchal structures reappeared in new clothes in a new setting. Instead of supervising an apparent 'rational self-organization of the market', the new power holders used their discretion to shape the rules of the game flexibly, according to their interests (Mattl 2003). Initiatives perceived as too radical or too politicized were no longer funded. Market pluralism was accepted only in those fields convenient to governmental institutions, thereby excluding, or marginalizing, the minority from access to funding or public posts. The right-wing government transformed Austrian corporatism, based on two partners on an equal footing (see Table 8.1), into an authoritarian system giving arbitrary power to politicians or higher public bureaucrats, who considered the state – within a limited time span and a given legal framework – as their property. It abandoned consensus politics and monopolized power in the hands of central government. 'Winner takes all' and 'Don't bite the hand that feeds you' have become the authoritarian *leitmotif*. Opposition and critique became increasingly difficult (Dimmel and Schmee 2005). This implied a radically conservative and anti-pluralist bias and created a culture adverse to social innovation and biased against initiatives that countered the short-term interests of dominant groups.

The positive outcome of these transformations was that some initiatives, like the *Association pour une Taxation des Transactions financières pour l'Aide aux Citoyens* (ATTAC) or newspaper projects like *Malmö* and *Augustin*, emerged as financially independent from the state. But the social-liberal market regulation of social policy implemented by the EU was challenged neither by the political parties nor by the actors in the third sector. On the contrary, it was still perceived as emancipatory and alternative to rigid national structures. Regarding re-nationalization via patrimonialism one could observe passive opposition that was reactive and focused on the procedural dimension of politics (Mattl 2003). Concerning the content innovations, there existed a certain nostalgia for the welfare past, but hardly any innovative ideas were proposed. The hegemonic consensus was able to reify capitalist market economies as the sole and natural social order (Jessop and Sum 2006). Changes in the socioeconomic order have not challenged the consensus; therefore, non-conformity in the political agenda is seen as unconstructive.

The Necessary Up-scaling of Social Innovations

What can be learned from history? First, it is difficult to achieve innovations which are concerned with both outcome and process at the same time. Whereas a more outcome-oriented and materialistic definition of social innovation was prevalent after the Second World War, increasing emphasis has been given to institutional aspects of social innovation since the 1990s. The deliberate integration of the two dimensions, the ultimate objective of integrated development, was limited to short time spans in the interwar period and at the beginning of the crisis of Fordism in the 1980s. At other times, the initiatives were either too localist and, therefore, lost their material dimension (their capacity to contribute to socioeconomic development) or too centralist and, therefore, despite becoming effective in the material dimension – as in the case of the Welfare State – retained processes that were paternalistic at best, and authoritarian at worst. Second, social innovations can occupy their appropriate place only if money, the market and property are repoliticized and re-embedded in the political economy. Only then can people shape development and make history and geography in diverse spaces, reappropriating political space to form their '*res publica*' (Novy and Leubolt 2005). Based on communication and cooperation between actors, people can transform themselves from recipients of funding to subjects of participatory planning, integrated area development and cooperative economy (Moulaert 2000; Albert 2004). Third, the dualism of top-down and bottom-up has to be revised, as has the dualism of state and civil society. The most creative bottom-up initiatives, be it after the First World War or in the 1980s, were supported by the local or, later, the national state. In the Austrian case, civil society is strongly linked to the state, and innovative autonomous strategies flourished when a government with political majority supported it: social mobilization has not been 'in stark contrast to the traditions of social reform' (Friedmann 1987, 83). *Red Vienna* and the Kreisky era are two examples which show that social reformism, social learning and social mobilization go hand in hand.

Civil society as a field of power is Janus-faced, as the experience of the 1990s showed (Novy 1996; Swyngedouw 2005). As Gramsci suggests, a context-sensible strategy must focus on the political in a broad sense, covering not only the state, but also civil society. It has to integrate elements of what he identifies as Eastern and Western strategies (1971, 238) – attending to experiments with diverse bottom-up initiatives as well as to the state's power to secure and institutionalize these experiments (cf. Poulantzas 2002). In this respect, it remains worthwhile to study *Red Vienna* as a laboratory of social innovation. And the experiences at the beginning of the 1980s should be re-studied with respect to their potential for alternative socioeconomic development through linking macroeconomic policies to bottom-up initiatives. In a nutshell, the best strategy for fostering social innovation today is a mode of governance that integrates democratic state-led social reformism with pluralist socially creative innovations from below.

References

Achs, O. (1993), 'Das Rote Wien und die Schule', in Öhlinger (ed.).

Agnoli, J. and Brückner, P. (1967), *Die Transformation der Demokratie* (Berlin: Voltaire).

Albert, M. (2004), *Parecon* (London: Verso) (originally published 2003).

Aly, G. (2005), *Hitlers Volksstaat* (Frankfurt: Fischer).

Aly, G. and Heim, S. (1993), *Vordenker der Vernichtung. Auschwitz und die deutschen Pläne für eine neue europäische Ordnung* (Frankfurt: Fischer).

Anderson, P. (1978), *Passages from Antiquity to Feudalism* (London: Verso).

Anderson, P. (1980), *Lineages of the Absolutist State* (London: Verso).

Becker, J. (2000), 'Verspätung und Avantgardismus. Zur Politischen Ökonomie des gesellschaftlichen Umbruchs in Österreich', *Kurswechsel* 15:4, 95.

Becker, J. and Novy, A. (1999), 'Divergence and Convergence of National and Local Regulation: the case of Austria and Vienna', *European Urban and Regional Studies* 6:2, 127–43.

Becker, J., Novy, A. and Redak, V. (1999), 'Austria between East and West: SRE Discussion Paper 69' (Vienna: Vienna University of Economics and Business Administration).

Berger, P. (2007), *Kurze Geschichte Österreichs im 20. Jahrhundert* (Wien: WUV).

Böck, S. (1993), 'Neue Menschen', in Öhlinger (ed.).

Bröckling, U., Krasmann, S. and Lemke, T. (eds) (2000), *Gouvernementalität der Gegenwart. Studien zur Ökonomisierung des Sozialen* (Frankfurt: Suhrkamp).

Cowen, M.P. and Shenton, B.W. (1996), *Doctrines of Development* (London: Routledge).

De Muro, P., Hamdouch, A., Cameron, S. and Moulaert, F. (2008). 'Combating Poverty in Europe and the Third World: social innovation in motion', in Drewe, Klein and Hulsbergen (eds).

Dimmel, N. and Schmee, J. (2005), 'Politische Kultur in Österreich 2000–2005', in Dimmel, N. and Schmee, J. (eds), *Politische Kultur in Österreich 2000–2005* (Wien: Promedia).

Drewe, P., Klein, J.-L. and Hulsbergen, E. (eds) (2008), *The Challenge of Social Innovation in Urban Revitalization* (Amsterdam: Techne Press).

Faßmann, H. (1995), 'Der Wandel der Bevölerungs- und Siedlungsstruktur in der Ersten Republik', in Talos (ed.).

Friedmann, J. (1987), *Planning in the Public Domain* (Princeton, NJ: Princeton University Press).

Friedmann, J. (1992), *Empowerment: the politics of alternative development* (Cambridge: Blackwell).

García, M. and Claver, N. (2003), 'Barcelona: governing coalitions, visitors, and the changing city center', in Hoffman, Fainstein and Judd (eds).

Gerhardter, G. (2000), 'Eigenständige Regionalentwicklung: Rumpelstilchen-Strategie und Akteurs-Management', in Roessler et al. (eds).

Gramsci, A. (1971), *Selections from the Prison Notebooks*, edited and translated by Qu. Hoare and G.N. Smith (London: Lawrence and Wishart).

Harvey, D. (1996), *Justice, Nature and the Geography of Difference* (London: Blackwell).

Hobsbawm, E. (1990), *Nations and Nationalism since 1780: programme, myth, reality* (Cambridge: Press Syndicate of the University of Cambridge).

Hoffman, L.M., Fainstein, S.S. and Judd, D.R. (eds) (2003), *Cities and Visitors* (Oxford: Blackwell).

Huffschmid, J. (2007), 'Die neoliberale Deformation Europas', *Blätter für Deutsche und Internationale Politik* 3, 307–19.

Jäger, J. (2003), 'Urban Land Rent Theory: a regulationist perspective', *International Journal of Urban and Regional Research* 27:2, 233–49.

Jessop, B. (1990), *State Theory: putting the capitalist state in its place* (University Park, PA: Pennsylvania State University Press).

Jessop, B. and Sum, N.-L. (2006), *Beyond the Regulation Approach: putting capitalist economies in their place* (Cheltenham: Edward Elgar).

Johnston, W.M. (1983), *The Austrian Mind* (London: University of California Press).

Leubolt, B., Novy, A. and Beinstein, B. (2007), *KATARSIS D1.5 Survey Paper: Governance and Democracy*, <http://katarsis.ncl.ac.uk/wp/wp1/ef5.html>, accessed 6 March 2008.

Maderthaner, W. (1993), 'Die österreichische Sozialdemokratie von 1919 bis 1934', in Öhlinger (ed.).

Mattl, S. (2003), 'Informelle Politik und flexible Institutionen. Eine verfrühte Bilanz der österreichischen "Wende-Politik" nach 2000', *Kurswechsel* 18:2, 9–18.

Melinz, G. (1999), 'Das "Rote Wien" als lokaler Sozialstaat: Möglichkeiten und Grenzen reformorientierter Kommunalpolitik', *Kurswechsel* 14:2, 17–27.

Moulaert, F. (2000), *Globalization and Integrated Area Development in European Cities* (Oxford: Oxford University Press).

Moulaert, F. and Cabaret, K. (2006), 'Planning, Networks and Power Relations: is democratic planning under capitalism possible?', *Planning Theory and Practice* 5:1, 51–70.

Moulaert, F., Martinelli, F., Gonzáles, S. and Swyngedouw, E. (2007), 'Introduction: social innovation and governance in European cities', *European Urban and Regional Studies* 14:3, 195–209.

Moulaert, F., Martinelli, F., Swyngedouw, E. and González, S. (2005), 'Towards Alternative Model(s) of Local Innovation', *Urban Studies* 42:11, 1969–90.

Moulaert, F. and Nussbaumer, J. (2005), 'The Social Region: beyond the territorial dynamics of the learning economy', *European Urban and Regional Studies* 12:1, 45–64.

Novy, K. (1993), *Klaus Novy – Beiträge zum Planungs- und Wohnungswesen* vol. 41, Beiträge zur Stadtforschung, Stadtentwicklung und Stadtgestaltung (Wien: Magistrat der Stadt Wien).

Novy, A. (1996), 'Zivilgesellschaft: Hoffnungsträger oder Trojanisches Pferd?', *Kurswechsel* 11:1, 26–38.

Novy, A. (2002), *Entwicklung gestalten. Gesellschaftsveränderungen in der Einen Welt, GEP* (Frankfurt/Wien: Brandes and Apsel/Südwind).

Novy, A. and Hammer, E. (2007), 'Radical Innovation in the Era of Liberal Governance: the case of Vienna', *European Urban and Regional Studies* 14:3, 211–34.

Novy, A., Lengauer, L. and Coimbra de Souza, D. (2007), 'Vienna in an Emerging Trans-national Region: socioeconomic development in the Central European region Centrope' (Vienna: Institute for the Environmental and Regional Development, Vienna University of Economics and Business Administration).

Novy, A. and Leubolt, B. (2005), 'Participatory Budgeting in Porto Alegre: social innovation and the dialectical relationship of state and civil society', *Urban Studies* 42:11, 2023–36.

Novy, A., Redak, V., Jäger, J. and Hamedinger, A. (2001), 'The End of Red Vienna: recent ruptures and continuities in urban governance', *European Urban and Regional Studies* 8:2, 131–44.

Oberhuber, F. (2005), 'Deliberation or "Mainstreaming"? Empirically researching the European convention', in Wodak and Chilton (eds).

Öhlinger, W. (ed.) (1993), *Das Rote Wien 1918–1934* (Wien: Museen der Stadt Wien).

Öhlinger, W. (1993), 'Im Spannungsfeld der Ersten Republik', in Öhlinger (ed.).

Peck, J. (2008), 'Remaking Laissez-faire', *Progress in Human Geography* 32:1, 3–43.

Pelinka, A. and Rosenberger, S. (2000), *Österreichische Politik: Grundlagen – Strukturen – Trends* (Wien: WUV-Universitätsverlag).

Pelizzari, A. (2001), *Die Ökonomisierung des Politischen. New Public Management und der neoliberale Angriff auf die öffentlichen Dienste* (Konstanz: UVK).

Pirhofer, G. (1993), 'Die Roten Burgen', in Öhlinger (ed.).

Pirker, R. and Stockhammer, E. (2006), 'Zur Aktualität austromarxistischen ökonomischen Denkens: Max Adler und Rudolf Hilferding', *Kurswechsel* 21:4, 27–36.

Polanyi, K. (1978), *The Great Transformation. Politische und ökonomische Ursprünge von Gesellschaften und Wirtschaftssystemen* (Frankfurt: Suhrkamp) (originally published 1944).

Poulantzas, N. (2002), *Staatstheorie. Politischer Überbau, Ideologie, Autoritärer Etatismus. Mit einer Einleitung von Alex Demirovic, Joachim Hirsch und Bob Jessop* (Hamburg: VSA) (originally published 1978).

Roessler, M., Schnee, R., Spitzy, C. and Stoik, C. (eds), *Gemeinwesenarbeit und Bürgerschaftliches Engagement* (Wien: OEGB Verlag).

Rohrmoser, A. (2000), 'Gemeinwesen als Strategie', in Roessler et al. (eds).

Schandl, F. and Schattauer, G. (1996), *Die Grünen in Österreich. Entwicklung und Konsolidierung einer politischen Kraft* (Wien: Promedia).

Schorske, C.E. (1982), *Wien. Geist und Gesellschaft im Fin de Siècle* (Frankfurt: Fischer).

Schumpeter, J.A. (1932), *Development*, <http://www.schumpeter.info/>, accessed 6 September 2007.

Sen, A. (1999), *Development as Freedom* (Oxford: Oxford University Press).

Stimmer, K. (2007), 'Stadtregulierungen vor 1945', *Perspektiven* 1:2, 7–15.

Stöhr, W.B. and Taylor, F.D.R. (eds) (1981), *Development from Above or Below? The dialectics of regional planning in developing countries* (Chichester: Wiley).

Stöhr, W.B. and Tödtling, F. (1978), 'Spatial Equity: some anti-theses to current regional development doctrine', *IIR – Discussion Paper* 3 (Wien: Wirtschaftsuniversität Wien).

Swyngedouw, E. (2005), 'Governance Innovation and the Citizen: the Janus face of governance-beyond-the-state', *Urban Studies* 42:11, 1991–2006.

Talos, E. (ed.) (1995), *Handbuch des politischen Systems Österreichs. Erste Republik 1918–1933* (Wien: Manz).

UNDP (2006), *Human Development Report 2006. Beyond Scarcity: power, poverty and the global water crisis* (Oxford: Oxford University Press).

Unger, B. (2006), 'Zählt der Austrokeynesianismus zur gesellschaftskritischen Ökonomie?', *Kurswechsel* 21:4, 66–78.

Vocelka, K. (2002), *Geschichte Österreichs. Kultur – Gesellschaft – Politik*. (München: Wilhelm Heyne).

Weber, M. (1980), *Wirtschaft und Gesellschaft. Grundriß der Verstehenden Soziologie* (5th edition, Tübingen: Mohr Siebeck) (originally published 1922).

Wodak, R. and Chilton, P. (eds) (2005), *A New Agenda in (Critical) Discourse Analysis: theory, methodology and interdisciplinarity* (Amsterdam: John Benjamins).

Chapter 9

Inclusive Places, Arts and Socially Creative Milieux

Isabel André, Eduardo Brito Henriques and Jorge Malheiros

Discussing the Concepts

Social Innovation as a Cultural Turn

Social innovation is an ambiguous concept with multiple meanings and vague contours. The expression has been increasingly used in the context of the restructuring of market relations driven by neoliberal orientations, as in the case of 'goal-based assessments', 'learning by doing', e-commerce, self-service distribution, private retirement plans and so on (Cloutier 2003; Alter 2000; Bassand 1986).

A quite different point of view, which is the one adopted in this text, relates social innovation to the development or provision of new responses to previously unsatisfied social needs, or even to previously unrecognized needs. These responses aim at promoting social inclusion by way of changing social relations, in this case through the empowerment of weak agents, or communities, and by reinforcing their relational capital (Hillier et al. 2004; Nussbaumer and Moulaert 2004; Klein and Harrison 2007).

> Social innovation – in both its product and process dimensions – is characterised by at least three forms of achievements, alone or in combination, accomplished through some form of collective action, as opposed to individual action: 1. it contributes to satisfy human needs not otherwise considered/satisfied; 2. it increases access rights (e.g. by political inclusiveness, redistributive policies, etc.); 3. it enhances human capabilities (e.g. by empowering particular social groups, increasing social capital, etc.). (Martinelli et al. 2003, 47–8)

Along similar lines, Yves Comeau (2004) considers social innovation as a broader concept than that applied to technological innovation, encompassing issues such as the reshaping of places, the emergence of new services and the development of new power relations.

In this sense, social innovation involves something out of the box, a new way of doing or thinking, a qualitative social change, an alternative to traditional procedures or even a break with the ways of the past. The reconstruction of social ties and structures inherent to social innovation involves risk and challenge. In fact, social innovation constitutes an effective threat against established order: 'L'innovation affronte l'institué, c'est-à dire qu'elle défait la tradition, elle dépasse la routine et elle défie les contraintes' [Innovation challenges the institutionalized, that is, innovation confronts tradition, overcomes routine and challenges the constraints] (Comeau 2004, 37).

The cultural challenge – linked to the reshaping of social ties and power relations – that is associated with social innovation, also implies the emergence of new governance models based on new ethics, solidarity, cooperation, reciprocity and tolerance.

> A plurality of social milieux with their own associations and collective spirits that derive from a variety of shared historical experiences, and common frames of reference, are what constitute contemporary urban societies. It becomes crucial for social innovation, if and how the plural structure is either causing fragmentation or producing a shared sphere of reciprocal responsibility and solidarity. (Gerometta et al. 2005, 2018)

Arts, Creativity and Socially Creative Milieux

> If we define culture as the creative elements of our existence – expressions of who we are, where we come from, and where we wish to go – then culture is the very thing which stores and carries those subjective, tradition-rich, value-laden institutional structures that undergird the sources of our well-being. (Jeannotte and Stanley 2002, 135)

In the context of the cultural turn brought about by social innovation, it is important to ascertain the role of creativity and the arts in the promotion of social innovation and in the construction of socially creative milieux. 'An inclusive civil society is not just a given. So, the remaining question is: how to create it? To reconstruct social relations at a local level would be real social innovation' (Gerometta et al. 2005, 2019).

Creativity can be seen as the human ability to produce new things, or create new situations, and is quite different from the notion of an innovative use of available resources and/or of known technologies. It deals with the human faculty of imagination. In writing, for instance, creativity consists of the ability to imagine new stories or use language in a personal and imaginative way. In the fine arts, for example painting, creativity can be expressed in the use of unexpected materials, in new combinations of colours and shapes, or through the ways in which something is done. When a child takes a piece of wood from the forest and transforms it into a toy, it is imagination and creativity that are driving the child.

Creativity is not exclusive to artistic production. Creativity occurs in all domains of human life. Even in scientific and technological knowledge, some dose of creativity is crucial: innovation depends much more on creativity than on memory (the ability to remember something or save information) or intelligence (the ability to learn, to know and to understand). In the arts, however, more than anywhere else, creativity is seen as a seminal ingredient in what is perceived as the qualitative evaluation of output, and is therefore highly appreciated and intensely cultivated. 'In modern Western art ... the rank of an artist's evaluation ... depends on how innovative he or she is considered to be' (Frey 2002, 363). Other qualities, such as technical skill, proficiency, or conformity to a given style or norm, are usually considered much less relevant.

Creativity is a value that artists gradually incorporate into their practices, encouraged by school, or by informal learning agents. As Caves (2000) has shown, this seems to be emphasized at art schools and conservatoires, where originality and expressiveness appear to be much more appreciated by teachers than any practical or theoretical knowledge. On this account, young artists are particularly impelled to develop and exercise those qualities. This is partly why artists are tempted to maintain an intense social life with their peers and why they tend to adhere to bohemian life styles: 'a continuous dialogue takes place to establish what are the major issues and new ideas – about art, philosophies and attitudes as well as techniques and materials. This is a key point in explaining the clustering of young artists' (Caves 2000, 26).

There is a natural component acting in creativity. Frey (2002, 368) expresses this idea by calling it 'personal motivation', but personal characteristics alone are insufficient to explain creativity. Creativity not only depends on biological and psychological determinants, but also, as we have suggested earlier, always needs some external source of inspiration. Keeping the relevance of those external or environmental factors in mind, it makes sense to regard creativity as a contextual effect (Drake 2003), and the creative milieux as simultaneously a product of, and a condition for, creativity. 'All societies need the symbolic resources which only culture produces. We need those strategic resources to make sense of our lives and to enable us to connect with one another and form productive collaborations' (Jeannotte and Stanley 2002, 138).

> Insofar as it constitutes an alternative to traditional ways of doing things, an avant-garde process and an unconventional response, social innovation is closely related to artistic production. ... Artistic expression entails recognition, provided that it refers to the essential symbols – collective signs – of belonging to groups, communities, cities, neighbourhoods. The real challenge for artistic expression in communication is to (re)construct identities that can be shared by a multitude of groups: neighbourhood arts and neighbourhood arts centres where a diversity of ethnicities, social and artistic groups all become anchored; arts that reflect a shared image of the city; educational projects that give access to artistic forms which short-circuit social fragmentation. (Nussbaumer and Moulaert 2004, 255)

Moreover, cultural resources, broadly defined, also constitute important driving forces in bringing about social innovation.

> Culture is to be understood ... as a 'voice' for deprived populations, communicating despair but also calling for consideration and respect. The – often spontaneous – development of specific infrastructures for alternative music, theatre or other artistic activities within deprived neighbourhoods shows this need for expressing contestation and a desire for change. (Moulaert et al. 2004, 231)

Two Socially Creative Places

Abundant literature has been produced in the last ten years on the issue of creative places (Scott 1997; Landry 2000; Hall 2000; Ley 2003; Florida 2002) and many urban plans have embraced this vision. Embedded in the 'new urban economy', the creative city 'movement' emphasizes the crucial role played by the cultural sector and by the creative industries, and also the importance of artists and of such factors as symbolic triggers, place marketing, cooperation, collective ambition, cultural diversity and social tolerance.

Are such creative milieux socially innovative? There is no simple answer. Some preconditions associated with the creative city approach, such as tolerance and cultural diversity, are also crucial to social innovation. However, some of its other features – such as public space theatricalization, or gentrification processes associated with 'artistic neighbourhoods' – are sources of social alienation and exclusion. An alternative view of urban development highlights the role of bottom–up processes in the mobilization of creative resources, as a crucial condition of local social innovation (Moulaert et al. 2005).

The cases presented in this chapter seek to promote the debate around the issue of socially creative milieux. Two short geographical explorations are pursued in order to understand the ties between the arts and social innovation in the production of socially creative milieux. Before presenting the roadmap to these explorations, it is important to stress that their ultimate underlying motivation stems from a concern with local development in the context of more or less global social innovation diffusion processes.

Inspired by the ALMOLIN model (Moulaert et al. 2005) and following the analytical framework on social innovation developed by Cloutier (2003) and André and Abreu (2006), a roadmap is developed to guide us on these two explorations. The first step deals with the context and aim of the initiatives; the remaining steps highlight five main constitutive dimensions:

- *from whom* – the agents of innovation
- *to whom* – the adopters
- *how* – the diffusion channels

- *constraints* – inertia and resistance factors
- *impacts* – effects of social innovation in terms of local development.

The agents of innovation – whether individual or collective – are those who introduce a new idea into a certain context at a given time. These agents can either introduce a novel 'invention', or import and adapt something from elsewhere. Creativity – a result of individual characteristics, but also related to participation in local, regional and international networks – is a key condition for invention or adaptation. It is also crucial in winning over the adopters, that is to say, in promoting the social appropriation of the innovation in question.

The adopters are key protagonists in the innovation processes. Without them, innovation would be nothing but an ephemeral good idea. The timing of the adoption processes varies in accordance with the characteristics of the potential adopters (age, gender, levels of schooling, professional activities, socialization process and so on), but it is largely determined by the structure of the milieux. In creative milieux, adoption tends to be quicker and more intense, and to involve unexpected adopters.

The diffusion channels consist of the resources deployed to promote the dissemination of the innovations. Civic participation and relational capital play a seemingly crucial role in strengthening the diffusion channels that are set in motion by the innovation agents and/or adopters, or even by other mediators such as the mass media. Diffusion channels can be either formal or informal, and strategically conceived or otherwise. Regardless of these distinctions, however, the more powerful they become, the better they are able to deal with social and cultural diversity.

The constraints act in the opposite direction from the waves generated by the diffusion channels. They are conservative forces; they refuse risk in order to defend the established order; in sum, they block innovation. Even in the most socially creative milieux, there are always some factors of inertia acting against innovation.

The impact of social innovation on the quality of places should not be reduced to their physical representation or landscape characteristics (city quarters, buildings, public spaces and suchlike). Place regeneration also involves the renewal of the cultural values and practices that shape social relations at a variety of different levels (family, neighbourhood, labour relations, citizenship and so on).

Developing Montemor-o-Novo through the Arts

Montemor-o-Novo (M-o-N) is a small town in Alentejo with a population of about nine thousand inhabitants in 2001, with approximately the same number residing in the municipality's rural areas. The town is connected to Évora (30 km) and Lisbon (100 km) by highway, making it very accessible.

Historically, M-o-N hosts a rural aristocracy, owning large farming estates in the area, and a rather poor rural proletariat. In the 1960s, the region's rural parishes lost over half of their population through migration abroad or to the Lisbon Metropolitan Area. Even the town saw a considerable number of inhabitants leave (5,636 residents in 1960 declining to 4,935 in 1970).

The deep social cleft that characterized this place until the 1970s and the intense exploitation to which its farmhands were subject ignited a strong opposition to Salazar's regime, masterly portrayed by José Saramago (Nobel Prize for Literature, 1998) in his novel *Levantado do Chão* [*Raised from the Floor*] (1980), the events of which take place in Lavre (a rural settlement in the M-o-N municipality). The opposition manifested itself in a number of ways, directly through labour strikes and political demonstrations (vigorously repressed by the police), and indirectly, via, for instance, associative movements and grassroots initiatives.

It was only after the 1970s – with the energetic dynamics shed upon this territory by the April 1974 Revolution and the Agrarian Reform in particular – that the town of Montemor developed significantly, evidenced by recent rises in population (6,408 in 1981, 7,056 in 1991, 8,766 in 2001).

The diversity of cultural initiatives currently promoted by non-government associations in M-o-N – from the older ones, such as the two philharmonic societies (one bringing together local landowners and other local influential people, the other one functioning as a leisure centre for the working class), to the more avant-garde projects in the fields of dance or sculpture – is, to a large extent, still an inheritance from the revolutionary period.

But these initiatives are not merely an inheritance; they are also a product of the social innovation that the town has been able to promote by combining top-down strategies (undertaken by the communist municipality) with bottom-up projects promoted by civil society organizations from varying ideological orientations. A number of projects in the forefront of cultural and artistic creation and promotion are being developed alongside traditional initiatives (such as folklore, handicraft, festivities and so on). Between one and the other, there is also room for adding value to memory.

The multiplicity and diversity of cultural projects in M-o-N are tied to the local municipal strategy. Indeed, during these last decades, the M-o-N communist mayor has favoured culture as the main strategic axis on which to build local development, and culture not solely of a more classical, heritage-related kind, but also linked to creativity. The municipality has actively supported initiatives by artists and cultural programmers, either directly or as mediator between local agents and national authorities. In addition to logistic and financial support, municipal policy became more proactive, evidenced, for example, in the establishment of several partnerships, in the intense promotional campaigns for local events and in the invitation extended five years ago to Rui Horta – a Portuguese choreographer of vast international renown – to start the Montemor-o-Novo Choreographic Centre, 'The Space of Time' [O Espaço do Tempo], granting him use of the Saudação Convent (sixteenth century) facilities, located within the castle walls.

From a set of projects linked to artistic education, the theatre, dance, visual arts and literature, we have chosen to highlight a specific entity and initiative:

the International Symposium on Terra(cotta) Sculpture, organized by the Cultural Association for the Arts and Communication – 'The Convent Workshops' [Oficinas do Convento].

Created in 1996, the Convent Workshops are located at the São Francisco Convent, a municipal property in which a series of ateliers, a photography studio, show rooms and an artists' residence were created in the sequence of several rehabilitation works. In addition to artistic creation, the Convent Workshops promoted a series of conferences – utilizing the castle, the convents and the river – which underlined a concern to reinforce, a sense of local identity, a sense of the place that is M-o-N through the medium of art.

The Workshop Director's personality is not unrelated to the Convent Workshops' dynamic output. Linked to the countless cultural actions of revolutionary nature which 'invaded' Portugal following April 1974, Vírginia Fróis arrived to teach at the Faculty of Fine Arts at the University of Lisbon. Her close relationship (geographic and political) to local authorities on the one hand, and her tight network of national and international contacts on the other, together with her personal commitment, have allowed her to turn the Convent Workshops into a space of veritable socio-cultural innovation.

The promotion for the International Symposium on Terra(cotta) Sculpture stands among the most interesting of the Convent Workshop initiatives. The theme of the first edition (1999) was '7 Suns 7 Moons', and aimed at bringing artistic creation into the region in the area of sculpture. The Symposium's organization issued invitations to various sculptors (both Portuguese and international) to bring along recognizably important and promotionally interesting works and activities. Symposium II (2000) followed along the lines of its predecessor, reinforcing the intention to heighten the interchange between high art and local know-how and based on an investigation and survey carried out among its main promoters, such as embassies, the Orient Foundation [Fundação Oriente] and various cultural *attachés*. The central idea was to transform the Symposium into a clearly international event, with the municipality as main promoter.

Symposium III (2001), contrary to previous editions which worked by invitation only, followed the more accessible model of a competition, becoming an internationally publicized event for Terra(cotta) sculptures. The Convent Workshops set a number of goals for the Symposium; one which stands out is the establishment/strengthening of the interaction between artistic creation and heritage, contemporary art and traditional technologies, and the various fields of knowledge (arts, humanities, social sciences, natural sciences).

I received support from many people including the art students, technicians, artists and others who were now, without knowing it, repeating the rituals performed by their ancestors, who erected the stones which today challenge our imagination. And it was the subconscious memory of that ritual that pushed us to achieve something which at times seemed impossible. (César Cornejo, participating sculptor, Peru)[1]

The theme of Symposium III was 'To inhabit'; participating artists were asked to create their projects using local resources – materials (paste, ceramics and raw earth) and traditional techniques (preparing clay, kilning, firing). The ensuing works were positioned at several sites of historical and cultural interest in the municipality, thus integrating with its heritage.

> I landed in the country of the colonizers. I started to understand the origin of a lot of habits, ways of thinking and obsessions. I felt like being in Brazil. Building a 'João de Barro's' nest (a South American bird) in the country that colonized where I come from had great meaning – a reconquest. The building place was a holy gift. I still bring with me the smell of the hills, the wind and the sun and moon. (Rosana Bortolin, participating sculptor, Brazil)

This project contributed to the learning of new sculptural techniques and promoted close relationships between professionals from different backgrounds and fields of knowledge – relationships which eluded the rigidity of power structures. The project gathered architects, sculptors, plastic artists, teachers, masons and students, all of whom formed groups to experiment in creating sculptures that might serve to enhance the town's public spaces. At the same time, the Symposium mobilized the local population (much of which was frequently present during the sculpting, maintaining a constant dialogue with the artists), and gave these men, women and children a daily opportunity to become acquainted with traditional skills as represented in the forms of modern sculpture. The project, therefore, celebrated their collective memory and inserted a heritage product/object into their present-day time – it became the creation of a new object backed by a whole collective memory. (Key features of the development of Montemor-o-Novo through arts are summarized in Table 9.1.)

Socially creative processes and initiatives occur likewise in different socio-spatial contexts and can even be trigged by ephemeral events promoted by mainstream agents as those engaged in the organization of *Porto 2001* – European Capital of Culture.

Playing Wozzeck in Aldoar

Originally, opera performances and their variants had targeted audiences that were socially diverse, but gradually, opera has become an art style that appeals only to certain segments of the educated elite. However, some contemporary music institutions and opera companies have incorporated the social dimension of erudite culture in addition to the obligatory aesthetic dimension. In these cases, classical

---

1   Translations of interviews are by the authors.

Table 9.1   Summary – developing Montemor-o-Novo through arts

Agents and triggers of innovation	• Municipality initiatives • Civil society organizations (from grassroots projects to those promoted by the old local aristocracy) • Partnerships involving the municipality and recognized cultural agents
Creators and adopters	• The artists • The new residents (middle-class intellectuals coming from Lisbon) • The older residents (democratization of cultural supply, recognition of informal knowledge and reassertion and reinforcement of self-esteem)
Diffusion channels	• Popular involvement • The role of charismatic actors • Local networks linked to global networks
Opportunities	• The role culture plays in the municipal development strategy • The involvement of recognized artists who mobilize high levels of social capital at national and international scales • The short travel time between M-o-N and Lisbon
Constraints – inertia and resistance factors	• Uncertain sustainability of local initiatives • Uncertain sources of financial support to the projects • Bureaucracy (e.g. the incapacity on the part of cultural grassroots organizations to draw on public support)
Impacts – effects of social innovation in terms of local development	• New social and economic dynamics (attraction of new residents) • Democratization and improvement of the local cultural supply • The example of M-o-N as 'good practice'

dance or opera are assumed to be relevant modes of communication, which can express, in a powerful and aesthetically pleasing way, situations, problems or ideas that touch all members of a community, including the lower classes. Within this framework, opera is not only about aesthetics and art, but also an instrument of social intervention and, eventually, of empowerment.

This social dimension of opera is clearly present in the work of the Birmingham Opera Company (BOC), as it is stressed by Graham Vick, its artistic director:

> Involving ourselves with our local communities, making the populations the realm of our work, the BOC may create a powerful means of communication. This is not, of course, what people expect from opera, but that is crucial and it is just that we are talking about. (Nicholson, Vick and BOC 2002, 31)

The work of BOC is well known and very highly regarded in the European artistic environment, and the Education Department (ED) of the Oporto *Casa da Música* [House of Music] was well aware of this when it decided to invite the BOC to explore possibilities of participation in *Porto 2001* – European Capital of Culture. In addition to the recognized quality of its performances, the explicit incorporation of the social dimension in the work culture of BOC fitted well with one of the aims of the *Porto 2001* initiative, i.e. to involve a large number of Oporto neighbourhoods in the project, including, as much as possible, the deprived quarters and their populations.

It was under this key principle of 'social intervention through art' that the Opera *Wozzeck*, by Alban Berg, was staged by Graham Vick in the old Central Eléctrica [Power Station] of Freixo and materialized as a sophisticated art performance through the participation of people living in a deprived neighbourhood of Oporto (sub-quarters of Aldoar and Fonte da Moura).[2] The institutional intention that *Porto 2001* involves the Oporto neighbourhoods in the event, and the social culture of BOC and ED's eagerness to bring together two apparent incompatibilities (the sophisticated and concert hall-placed art of opera and a neighbourhood where opera was an unknown word) were the elements that made the design, implementation and success of this project possible.

But how were the two *Porto 2001 Wozzeck* Opera performances built up? And what were the effects that really enable us to read Aldoar as a temporarily creative place?

After the first contacts between the ED of Casa da Música and the directors of BOC, the latter proposed involving about a hundred residents of a deprived neighbourhood with no experience of sophisticated art in the public performances of *Wozzeck* in Oporto. This meant that the professionals of BOC would mix with the Aldoar residents, in both the rehearsals and the two Opera recitals (Ralha 2002, 22–3).

The choice of Aldoar and Fonte da Moura from among several deprived neighbourhoods of Oporto arose from a mix of elements, but not from specific issues related to any living art traditions[3] or to any social movements more active here than in other areas of the city. On the one hand, these neighbourhoods were not involved in the European Capital of Culture initiative; on the other, they share the ingredients that make social intervention through opera so challenging: i) stigmatized districts with very bad images in the city, and frequently associated with crime; ii) lack of previous contact between the resident population and the art

---

2 The Opera, *Wozzeck*, was composed by Alban Berg between 1914 and 1920, and was based on the drama *Woyzeck* by Georg Büchner (1836–37). It relates the predictability of exploitation and privation for poor people. In a military setting, Wozzeck – an all-service soldier – is often insulted by the Captain, subjected to medical experimentation by an army doctor and, finally, betrayed by his wife Marie.

3 Although experiences of folk theatre organized by local associations had taken place in the past, these are practically dead nowadays. The only significant performative movement corresponds to folk marches that take place in June, during the S. João feasts. However, this is common to several other neighbourhoods of Oporto (interview with Suzana Ralha, from the DE of Casa da Música, January 2005).

of opera; and iii) the apparent lack of basic relevant competences within the local population.

In September 2000, after a few visits to Oporto, the ED of Casa da Música and the BOC confirmed Aldoar and Quinta da Moura as the Project sites, and secured the agreement of the sub-local authorities (Junta de Freguesia). However, this institutional cooperation was clearly insufficient to gain the support of local people and involve them in a project centred on a 'distant' kind of art, considered, *a priori*, as disagreeable and irrelevant. In addition, people living in such social quarters are frequently suspicious of this kind of project – of initiatives that seem to appeal to their involvement and participation, but turn out to be superficial actions, quickly forsaking the neighbourhoods and leaving the people feeling deceived time after time.[4]

Within this context, the first three months (October to December 2000) aimed to establish a trusting relationship between the professionals of the ED of Casa da Música and the local population. The ED's presence in the quarter was regular and frequent, but before both opera in general and *Wozzeck* in particular were presented to the residents, it was necessary to develop a common language of communication, to assure them that the project was to be taken through to the end, that their involvement was crucial and that all Oporto residents as well as the press would talk positively about the neighbourhoods. An important element in achieving the latter goal was a cooperation agreement established with a television broadcast company, which covered the whole project and has become a frequent presence in the quarters.

In January 2001, the direct work with the local residents started. This was the moment chosen to make both the idea of opera and *Wozzeck* itself appear in the process. The people of these quarters have negative feelings towards opera because it is unknown to them and they do not understand it; but they do like stories and they do like music. Through a step-by-step process of deconstruction of opera stories, professionals of the ED began to enthuse the local population about opera and its possibilities. The story of *Wozzeck* was presented to the communities, who went on to discuss the roles of the different characters (the soldier, his wife, the captain ...) and their symbolic meanings.

By February, the number of people interested in participating in the performances of *Wozzeck* reached 90. The project took a further step forward by requiring a more pro-active attitude from the population. A March Carnival contest was organized, requiring all participants to be dressed as *Wozzeck* characters. These costumes were made by the participants themselves (with financial support from *Porto 2001*), after discussion with the ED professionals working in the neighbourhoods. The competition judges, which included a famous national rock star (Pedro Abrunhosa) who visited the neighbourhood and directly contacted the residents, chose five winners; these were awarded a visit to Birmingham, including, naturally, a reception at the BOC.

---

4   Interview with Suzana Ralha, from the ED of Casa da Música, January 2005.

At this stage of the process (March 2001), three fundamental achievements had been realized: i) a relationship of trust had developed between the ED of Casa da Música and the local population; ii) the local population had embraced the project and assumed a more pro-active participation; and iii) the quarter had started to open up to the outside and its image had begun to be re-created. The increasing and ever-changing trade-off between these neighbourhoods and the outside city (and country) was visible in the presence of non-locals in Aldoar (people of the ED, Pedro Abrunhosa, other professionals of the artistic scene and so on), in the visits of the local residents to other places (Birmingham, opera performances in Oporto theatres and suchlike) and in the news spread about Aldoar within the framework of the *Wozzeck* project. According to the professionals associated with the initiative, the media played a very relevant role, helping to attract the local population and to diffuse a changing image of the neighbourhoods. There was also an evolution in the perspective of those journalists who followed the project closely: from a journalism marked by stereotypes (the difficulties in involving this 'kind' of population in the project, the social problems of the neighbourhoods and so on) towards a more comprehensive and constructive journalism, which stressed the merits of the project to both the populations (in terms of self-esteem, empowerment and the like) and the professionals and artists involved (Pereira, 2002: 54–6).

The professionals and opera singers of BOC arrived in Oporto in March and for ten days they worked together with the professionals of the ED of Casa da Música and the 108 inhabitants of Aldoar and Fonte da Moura who had decided to participate in the performances. The work was very intense and was only possible owing to the previous steps jointly taken by the local population and the Casa da Música professionals. The relative absence of in-group tensions in this phase of the project, usually common in collective experiences involving very different people, is good proof of the efficacy of the path followed in the previous months. In fact, it is important to remember that tensions do exist between men and women, old and young, drug addicts and locals and residents of two different sub-quarters in this neighbourhood. To overcome them and get the different elements working together was an achievement of this project that may last, contributing to an improvement in community relations; that is, to increasing the local sense of place.

The two performances of *Wozzeck* staged by Graham Vick and sung by BOC finally took place in early April in the Power Station of Freixo, situated outside Aldoar. They were a success on many levels: they were sold out; they achieved a high artistic and aesthetic quality; they were prepared in an unconventional way, involving collaboration between the public, the BOC singers and the Aldoar participants in the same venue, breaking up the traditional separations between socio-professional groups that traditionally do not share the same spaces and the same kinds of art. (Key aspects of the *Wozzeck* project are summarized in Table 9.2.)

Despite the success of the whole process, Aldoar as a creative place, even of a temporary type, is still a rather limited exercise. The process was entirely induced from outside and, despite its contribution to the changing image of the neighbourhoods and the progressive involvement of the local population and their presence in the core of the initiative, the final event did not take place in Aldoar itself (it happened in Central do Freixo). The logic of the process is based on the commitment of the local population to ideas and events designed by others. However, two follow ups to the *Wozzeck* performances show how the whole initiative has both increased the sense of community in Aldoar and Fonte da Moura and contributed to the development of the sense of a creative place.

In June 2001, two months after the performances of BOC at Freixo, some members of the local community approached an official of the ED of Casa da Música to ask about the possibility of organizing some concerts, eventually of erudite music, inside the quarter. This would bring people from the middle-high and high classes to the quarter and would open it to the outside, continuing the process that had been started with *Wozzeck*.

After a period of indecision among the inhabitants who proposed the idea, because of issues concerning logistics and especially security, three concerts – classical music and percussion – took place in Aldoar and Fonte da Moura in the summer of 2001.

The question of security cannot be considered a mere detail. Assurance of place security was one more step in the process of collective empowerment occurring in these neighbourhoods. Making it a condition that no specific police operation would be set up for the concerts, the local population itself took care of the performances' security. The concerts took place and were a success; and no criminal events were recorded.

After the summer, a new project was initiated, this time entirely conceived and performed by the local population with key contributions from several art professionals and the supervision of the ED of Casa da Música. In November 2001, a process of collective writing took place, involving a group of inhabitants together with experts in running writing workshops.[5] In December, this collective work materialized in the form of the libretto, *Demolition – the story you are about to see*. The story is centred on a building that represents the society of the quarter and its complex relationships. The main focus is a fight between the inhabitants of the building and the powers (the porter of the building and the public authorities, as represented by the mayor) who want to demolish it and use the space to build a shopping mall or a bank. In the end, the inhabitants lose and the building is effectively demolished, reflecting the inherent belief in the local population (who view themselves as the residents of the building) that this is the most plausible outcome in a story where working-class people fight against a much stronger power. The metaphor of *Demolition* is also connected to the reality of the Aldoar neighbourhood at the time; a re-housing process was taking place, involving demolition of shacks and the building of new homes.

---

5 The writing workshops were directed by the poet and dramaturge, Regina Guimarães.

Table 9.2   Summary – playing *Wozzeck* in Aldoar

Agents and triggers of innovation	• *Porto 2001* – European Capital of Culture • Education Department of the Oporto Casa da Música • BOC – Birmingham Opera Company
Adopters	• A deprived inner-city neighbourhood and its population
Diffusion channels	• Participation of people living in a deprived neighbourhood of Oporto in a sophisticated art performance • Cooperation agreement established with the media, including a television broadcast company
Constraints – inertia and resistance factors	• Initial resistance of the local population – difficulties in terms of communication • Tensions between men and women, between youngsters and older residents, and between drug addicts and other local residents in two different parts of the neighbourhood • Dependency upon public support • Security problems in the neighbourhood
Opportunities	• A political goal, expressed by the organizers of a prestigious European event, of linking culture and social initiatives • The socially creative project of Birmingham Opera Company
Impacts – effects of social innovation in terms of local development	• Regeneration of the social image of the neighbourhood • Reinforcement of the local social capital • Improvement of community relations • Opening up of the neighbourhood to the outside • Empowerment of the local community • Mobilization of artistic resources in order to identify and communicate local problems

However, the collective writing process was only one of several dimensions in popular involvement in the process of constructing the 'Garage Opera' *Demolition*. Residents also participated in the construction of musical instruments, they learnt and practised rhythms and choreographies and they participated in the rehearsals. Finally at the end of February/beginning of March 2002, in the semi-built premises of the future Casa da Música do Porto, the show *Demolition* was performed with the participation of 32 singers (29 non-professionals from Aldoar and Fonte da Moura and three professionals – a mezzo-soprano, a tenor and a baritone). The whole project was directed by Suzana Ralha from the ED, with music and lyrics written by professionals (as a result of a process of collective work with the local population).

Revising the Concept of Socially Creative Milieux: Final Remarks

The two geographical explorations presented here show how the implementation of formally de-contextualized artistic events can represent a process of social innovation, capable of empowering populations living in peripheral places and leading them to express their voices through a dialogue with artists and art producers and especially through participation in collective art productions (sculpture and opera). However, this was only possible because re-contextualization occurred and because the International Symposium on Terra(cotta) Sculpture, and the operas *Wozzeck* and *Demolition* were embedded in their places of production and development: Montemor-o-Novo and Aldoar. The specific cultural ambiance of M-o-N and the pro-active and revolutionary tradition of its local government contribute to the openness to new art experiences that stress the cultural dialogue between locals, artists and other actors, bringing together traditional craft skills and modern sculpture in the form of locally meaningful products. The rough social character of the Aldoar neighbourhood, together with its stigmatized image and its conflicts over housing, are paralleled in *Wozzeck*, as a BOC production, and totally expressed in the message and production process of the Garage Opera *Demolition*.

If these places were crucial to the development of the two cultural initiatives, their construction as creative milieux results from the process of using art resources (artists, plays, materials and so on) to enhance local mobilization and local expression of wishes and symbols. This reveals the relevant role of local policy options (of the M-o-N municipality and of *Porto 2001* – European Capital of Culture) in terms of integral development and inclusion through the arts, and the role of cultural agents (the Cultural Association for the Arts and Communication in M-o-N: the ED of Casa da Música in Porto, Birmingham Opera Company) and non-local artists, acting as both project stimulators and agglutination catalysts. The presence of these exogenous elements has contributed to the stimulation of diversity – a condition for social innovation – within a framework marked by actions aimed at increasing reciprocal responsibility (collective productions involving local residents; conditions to produce cultural events in terms of accessibility and security; and the like) and overcoming tensions and potential conflicts between the various groups of social actors (artists and non-artists, locals and non-locals, men and women and so on). This has been crucial in preventing the fragmentation of these increasingly plural structures, reinforcing instead their solidarity and inclusion potential.

All in all, the two examples presented suggest that while socially creative milieux typically have strong identities and a certain degree of resilience, they are also flexible enough to be able to change when subject to the action of combined endogenous and exogenous social innovations. In a word, they are typically 'plastic' in the physical sense of the concept, e.g. the object that can be shaped and reshaped without losing its coherence and its meaning (Lambert and Rezsöhazy 2004).

The concept of plasticity presented by Dominique Lambert handsomely sums up the crucial characteristics of socially creative milieux: they should be both reasonably ordered and sufficiently flexible to be reshaped without losing their identity, that is, without fragmentation.

Thus, in terms of their own plasticity, socially creative milieux seem to combine: i) diversity, in the sense that they are multicultural; ii) tolerance, in the sense that they embrace the risk of doing something new; and iii) active democracy, in the sense that citizens are encouraged to participate. The cases of Montemor-o-Novo and Aldoar provide good examples and illustrations of these crucial conditions (see Table 9.3).

Table 9.3   The plasticity of socially creative milieux: Montemor-o-Novo and Aldoar

	Developing Montemor-o-Novo through the arts	**Playing *Wozzeck* in Aldoar**
Cultural diversity	• Longstanding rural aristocratic culture • Revolutionary culture • Central role played by recognized contemporary artists from various art fields	• Ethnic mix among the artists • Audience made up of old and new elites, as well as relatives and neighbours of the participants in the opera • Heterogeneous local population – different age and gender groups, drug-addicts, sub-local communities
Risk and tolerance	• Acceptance of paradoxical objects in the public space	• Population subject to exclusion or at risk of exclusion participating in an elitist activity • Bringing together apparently incompatible universes (the world of opera and the deprived neighbourhood)
Active democracy	• Involvement of the local residents in artistic activities • Reinforcement of the self-esteem of the local populations • Social recognition of traditional skills and crafts	• Gradual involvement of excluded groups • Reinforcement of the self-esteem of the local populations • Relations between different groups

Despite the plasticity and inclusive nature of these socially creative milieux, some caveats remain and require further discussion. On the one hand, the leading roles played by non-local actors point to the relevance of 'external detonators' in social innovation processes taking place in peripheral areas with excluded populations, even if a bottom-up appropriation occurs at a later stage. On the other hand, the sustainability of the two creative milieux presented here is largely dependent, as in many other cases, on public resources and short-term policy strategies.

References

Alter, N. (2000), *L'innovation ordinaire* (Paris: PUF).

André, I. and Abreu, A. (2006), 'Dimensões e espaços da inovação social', *Finisterra* XLI:81, 121–41.

Bassand, M. (1986), *Innovation et Changement Social* (Paris: Presses Polytechniques et Universitaires Romandes).

Caves, R.E. (2000), *Creative Industries: contracts between art and commerce* (Oxford: Harvard University Press).

Cloutier, J. (2003), *Qu'est-ce que l'innovation sociale?* (Montreal: Cahiers du CRISES, Collection Études théoriques ET0314).

Comeau, Y. (2004), 'Les contributions des sociologies de l'innovation à l'étude du changement social', *Actes du Colloque Innovations Sociales et Transformations des Conditions de Vie*, 29–41 (Montreal: Cahiers du CRISES, Collection Études théoriques ET0418).

Drake, G. (2003), 'This Place Gives Me Space: place and creativity in creative industries', *Geoforum* 34, 511–24.

Florida, R. (2002), *The Rise of the Creative Class: and how it's transforming work, leisure, and everyday life* (New York: Basic Books).

Frey, B. (2002), 'Creativity, Government, and the Arts', *The Economist* 150:4, 363–76.

Gerometta, J., Haussermann, H. and Longo, G. (2005), 'Social Innovation and Civil Society in Urban Governance: strategies for an inclusive city', *Urban Studies* 42:11, 2007–21.

Hall, P. (2000), 'Creative Cities and Economic Development', *Urban Studies* 37:4, 639–49.

Hillier, J., Moulaert, F. and Nussbaumer, J. (2004), 'Trois essais sur le rôle de l'innovation sociale dans le développement spatial', *Géographie, Economie, Sociétés* 6:2, 129–52.

Jeannotte, M.S. and Stanley, D. (2002), 'How Will We Live Together?', *Canadian Journal of Communication* 27, 133–9.

Klein, J.-L. and Harrison, D. (eds) (2007), *L'innovation Sociale: Emergence et effets sur la transformation des sociétés* (Québec: Presses de l'Université du Québec).

Lambert, D. and Rezsöhazy, R. (2004), *Comment les pattes viennent au serpent. Essai sur l'étonnante plasticité du vivant* (Paris: Editions Flammarion).

Landry, C. (2000), *The Creative City: a toolkit for urban innovators* (London: Earthscan).

Ley, D. (2003), 'Artists, Aestheticisation and the Field of Gentrification', *Urban Studies* 40:12, 2527–44.

Martinelli, F., Moulaert, F., Swyngedouw, E. and Ailenei, O. (2003), *Social Innovation, Governance and Community Building – Singocom – Scientific Periodic Progress Report Month 18*, April 2003, <http://users.skynet.be/bk368453/singocom/ index2.html>, accessed 5 July 2007.

Moulaert, F., Demuynck, H. and Nussbaumer, J. (2004), 'Urban Renaissance: from physical beautification to social empowerment. Lessons from Bruges – Cultural Capital of Europe 2002', *City* 8:2, 229–35.

Moulaert, F., Martinelli, F., Swyngedouw, E. and Gonzalez, S. (2005), 'Towards Alternative Model(s) of Local Innovation', *Urban Studies* 42:11, 1969–90.

Nicholson, J., Vick, G. and BOC (2002), 'Birmingham Opera Company, Anatomia da Nova Ópera', in Pereira (ed.).

Nussbaumer, J. and Moulaert, F. (2004), 'Integrated Area Development and Social Innovation in European Cities: a cultural focus', *City* 8:2, 249–57.

Pereira, R. (ed.) (2002), *Construção – Demolição – Obras de Arte – Viagem ao interior de um projecto artístico* (Porto: Campo das Letras).

Pereira, R. (2002), 'Reportagem do "inimaginário": os media e a "excepção" de Aldoar', in Pereira (ed.).

Ralha, S. (2002), 'Vozes um pouco mais livres e mais humanas', in Pereira (ed.).

Saramago, J. (1980), *Levantado do Chão* (Lisbon: Caminho).

Scott, A.J. (1997), 'The Cultural Economy of Cities', *International Journal of Urban and Regional Research* 21:2, 323–40.

Index

A Postcapitalist Politics 26
Aldoar (Oporto)
    *Demolition – the story you are about to
      see* 161–2, 163
    Wozzeck opera 156–62, 163–4
ALMOLIN (Alternative Model for Local
    Innovation) 25, 152–3
arts
    Montemor-o-Novo (Portugal) 153–62
    social innovation 14–15, 150–2
Austria
    Austromarxists 134
    Central Bank 136
    civil society and social innovation 133,
      143
    corporatism 135, 142
    economic restructuring 138
    Fordism 137, 143
    governance of scale 131–47
    Keynesian Welfare State 135–6, 136–7
    Nazi occupation 134
    Public Employment Service 139–40,
      141
    Red Vienna project 5, 133–6, 143
    socially creative strategies 136–42
    social innovation 5–6, 131–47
      upscaling 143
    social reformism (1950s–1970s) 134–6
    welfare via top-down policies 132–6
    Zentrum fur Soziale Innovation 139

banks and microfinance 53
Barcelona
    Community Diagnostic document 93
    Trinitat Nova initiative 92–5
beneficiaries of social solidarity finance 46
Berg, Alban 158
Berlin
    business registrations/cancellations
      105–6
    corporate demography 105–6

ethnic economies 104–8
labour market 104–5
migrant entrepreneurship 101–2,
    102–14
migrants 4
non-governmental organizations 110
social innovation 101–2, 102–14
Turkish population 104–7
Vietnamese population 105–8
Birmingham Opera Company (BOC) 157,
    159–60, 163
'black success' in United States 103
BOC *see* Birmingham Opera Company
'bottom-up' institutions, social innovation
    18
Britain
    deprived neighbourhoods 82–3
    ethnic economies 103
business registrations/cancellations, Berlin
    105–6

capital mobilisation, microfinance 50–1
Casa de Musica (Oporto) 157–8, 160–2
Catacumba slum (Rio de Janeiro) 126
Catalonia community initiatives 90 ???
CGAP *see* Consultative Group to Assist
    the Poor
civil society and governance 66–70, 70–2
'commons', concept in community
    economies 34
Community Development Plan in Spain 90
Community Diagnostic document
    (Barcelona) 93
community economies
    distributive justice 29–30
    governance 31–2
    innovation dynamics 35
    innovative communities 32–4
    Latrobe Valley (Australia) 33
    marginalization 35
    Pioneer Valley (Massachusetts) 33

social innovation 25–34, 34–6, 35
Community Economies Collective 32
Community Economies project 25, 27, 30, 36
community empowerment 3–4
Consultative Group to Assist the Poor (CGAP) 42, 43
'consumption' concept in community economies 34
'content' dimension in social solidarity finance 45
Contrat de Ville 84
Conveni de Barri contract 93
'Convent Workshops' in Montemor-o-Novo (Portugal) 155–6
corporate demography in Berlin 105–6
'Corporate Social Responsibility' 11
corporatism, Austria 135, 142
creative milieux, socially innovative 149–66
creativity and social innovation 6, 14–15, 150–2, 153
culture and social innovation 149–52

Dallinger, Alfred 138
Demolition – the story you are about to see 161–2, 163
deprived neighbourhoods
    Britain 82–3
    regeneration 84–6
    United States 82–3
Développement Social des Quartiers policy 84
disintegration of urban areas 16–17
'*dispositifs*' in governance 68, 74
distributive justice in community economies 29–30
diversity in community economies 26–9
Durkheim, Emile 12–13

Eco-Neighbourhood Trinitat Nova programme 94
economics
    restructuring in Austria 138
    social innovation 18
    strategies 1
'empowerment dimension' in social solidarity finance 45

entrepreneurship
    Turkish population in Berlin 106–7
    Vietnamese population in Berlin 107–8
EQUAL Programmes (European Union) 108–9
ethnic economies
    Berlin 104–8
    Britain 103
    entrepreneurship 103
    'ethnic groups' 103
ethnic entrepreneurship
    Europe 102–4
    European Union 109
    United States 102–4
    urban development 112
'ethnic groups' economies 103
'ethnomarketing' 111
Ethnotrade magazine 111
EU *see* European Union
Europe
    ethnic entrepreneurship 102–4
    integrated area initiatives 84–5
    microfinance 40
European Framework Programme 3
European Union (EU)
    EQUAL Programmes 108–9
    ethnic entrepreneurship 109
    Framework for Action for Sustainable Urban Development 85
    governance, 64, 65, 141–2
    microfinance 40–1
    migrant populations 111
    Quartiers en Crise programme 90
    URBAN Community Initiative 85
    Urban Pilot Programme 85
excluded populations and social solidarity finance 46

'familial solidarity economy' of slum dwellers 123–7
festivalization and urban policies 110–11
Ford, Henry 14
Fordism
    Austria 137, 143
    Gramscian perspective 67–8, 70
Forschungs- und Beratungsstelle Arbeitswelt (FORBA) 139
'founding act' in Latin America 115

France
   Contrat de Ville 84
   Développment Social des Quartiers
      policy 84
   integrated area initiatives 84
Franklin, Benjamin 3, 14
funding innovation in social solidarity
      finance 46–7

'German Austria' 132
governance
   Argentina 31–2
   Austria and European Union 141–2
   Austrian scale 131–47
   'beyond-the-state' 4, 65–6, 71–5
   changing regime 70–2
   civil society 66–70
   community economies 31–2
   '*dispositifs*' 68, 74
   'glocalization' 73
   non-governmental organizations 63–4
   social innovation 63–5
   state 66–70
governance-beyond-the-state
   civil society 4, 71–2
   European Union 65
   networked associations 65–6
   power politics 72–3
governmentality 66–70
Grameen Bank (Bangledesh) 40
Gramsci, Antonio 67–8, 70
Guapore Housing complex (Rio de Janeiro)
      122–3, 127

Habsburg Empire 132
Heider, Bernhard 111
Horta, Rui 154
human capital and social solidarity finance
   47

IAD *see* Integrated Area Development
IHK (Chamber of Industry and Commerce)
      108–11
IMF *see* International Monetary Fund
innovation and communities 32–5
institutions and ethic entrepreneurship
      108–11

Integrated Area Development (IAD) 3, 15,
      17–19, 87
integrated area initiatives 17–18, 87
   Europe 84–5
   France 84
International Monetary Fund (IMF) 64,
   73
International Symposium on Terra(cotta)
      Sculpture 6, 155

Keynesian welfare state
   Austria 135–6, 136–7
   Gramsci perspective 67, 72
KRAX (urban ruptures) 16
Kreisky, Bruno 136, 143

labour market
   Berlin 104–5
   Latin America 117–20
Lambert, Dominique 163
Latin America
   'founding act' of land occupation 115
   labour market 117–20
   location preferences of slum dwellers
      117–20, 120–7
   'proximity economics' 119
   real estate markets 117–18
   slum dwellers 117–20
   social innovation 5, 115–16, 117–30,
      127–8
   urban territory 120–7
Latrobe Valley (Australia) community
      economies 33
Levantado do Chão (novel) 154
location preferences of slum dwellers in
      Latin America 117–20, 120–7
Luther King, Martin 14

Malmo newspaper 142
management science and social innovation
   14
marginalization
   community economies 35
   social innovation 35
marketization and urban policies 110–11
Marx, Karl 14, 66
MFIs *see* microfinance institutions
microcredit 39, 40, 41

microfinance
    banks 53
    capital mobilisation 50–1
    definition 39–43
    Europe 40
    European Union 40–1
    non-governmental organizations 43
    'pre-banking' 44–5
    social exclusion 51–4
    social solidarity finance 45–51, 51–5
    sustainability 50
microfinance institutions (MFIs)
    principles 41–3
    profitability 50
    social performance 43–5
    social relations 49
    social solidarity 47–8
migrants
    Berlin 4
    community economies 34
    entrepreneurship in Berlin 101–2,
        102–14
    social innovation 111–12
Mondragón Cooperative Corporation 30
Montemor-o-Novo (Portugal)
    arts 153–62
    'Convent Workshops' 155–6
Multipartite Social Contract (MSC) 54

Nazi occupation, Austria 134–5
'necessity' concept in community
    economies 34
needs satisfaction 2–3
neighbourhood regeneration
    local democracy 86–8
    social innovation 82–6, 86–8
    Spain 88–92
*Neighbourhood Renewal Operation of
    Madrid* 89
neighbourhoods 4–6
Neuköllner Bürgerstiftung (Civic
    Foundation of Neuköln) 110
non-governmental organizations (NGOs)
    Berlin 110
    governance 63–4
    governance-beyond-the-state 72–3
    microfinance 43
    poverty 55

slum transformation 121
    social exclusion 53–4
    social innovation 29
Nova Brasilia slum (Rio de Janeiro) 126

Oporto (*Casa de Musica*) 157–8, 160–2
'organized proximity' 120
*Österreichische Arbeitsgemeinschaft für
    Regionalentwicklung* (OAR) 139
*Österreichische Studien- und
    BeratungsGesellschaft* (OSB) 139
Ouseburn trust (Newcastle upon Tyne) 31

participation and social solidarity finance
    46
Pioneer Valley (Massachusetts) community
    economies 33
Polanyi, Karl 133
political science and social innovation
    15–16
Porte Alegre (Brazil) community
    economies 32–3
Porto 2001 – European Capital of Culture
    6, 156, 157–9, 163
power and governance-beyond-the-state
    72–4
'pre-banking' and microfinance 44–5
'process dimension' in social solidarity
    finance 45
'proximity economics' in Latin America
    119
public administration and social innovation
    15–16
Public Employment Service in Austria
    139–40, 141
Public-private partnership 1

Quartiersmanagement (district
    management) 109, 111

real estate markets in Latin America
    117–18
Red Vienna project (Austria) 5, 133–6, 143
(re)making of social space 6–8
residential mobility of slum dwellers 124
Rio de Janeiro
    Catacumba slum 126–7
    Guaporé Housing complex 122–3

Nova Brasilia slum 126
slum dwellers 119
Vidigal slum 124–5
Rosanvallon, Pierre 13

Saramago, José 154
Schader Foundation report 109
Schumpeter, Joseph 2–4, 12–13
SINGOCOM (Social Innovation in local
    Community Governance) 25, 27–8,
    29, 31, 86
slum dwellers
    'familial solidarity economy' 123
    Latin America 117–20
    location preferences 120–7
    residential mobility 124
    Rio de Janeiro 119
    Sao Paulo 119
slum transformation and non-governmental
    organizations 121
Social Enterprise Sunderland project 29
social exclusion 51–4, 81–2
social innovation
    ALMOLIN model 152–3
    analysis 7–8
    arts 14–15, 150–2
    Austria 5–6, 131–47
    Austria and civic society 133
    basic needs 17–18
    Berlin 101–2, 101–14
    'bottom-up' institutions 18
    community economies 25–34, 34–6
    concept 1–2, 11–12
    creative milieux 149–66
    creativity 6, 14–15, 150–2, 153
    culture 149–52
    economy 18
    empowerment 18–19, 44, 55, 64, 82,
        86–7
    ethnic economies 112
    governance 63–5
    history 12–14
    innovation dynamics 35
    integrated area development 3, 18–9,
        29
    Latin America 5, 115–16, 117–30,
        127–8
    management science 14

marginalization 35
migrants 111–12
neighbourhood regeneration 82–8,
    86–8, 88–92
non-governmental organizations 29
political science 15–16
public administration 15–16
(re)making of social space 6–8
social relations 45–7, 48–50, 87, 102,
    149–50, 153
social science 14–16
Spain 81–2, 82–100
territorial development 15, 16–21
Social Performance Indicators (SPIs) 43
social performance and microfinance
    institutions 43–5
social reformism in Austria (1950s–1970s)
    134–6
'Social region' model 15
social relations and territorial development
    19–21
social science and social innovation 14–16
social solidarity finance
    'content' dimension 45
    efficiency 49
    'empowerment dimension' 45
    excluded population 46
    funding innovation 46–7
    human capital 47
    microfinance 45–51, 51–5
    participation 46
    'process dimension' 45
    proximity to beneficiaries 46
    social relations 48–51
'Sociologist as an Artist' 14, 15
Spain
    Community Development Plan 90
    Integrated Area Development 87
    neighbourhood regeneration 88–92
    Rehabilitation Plans 90
    social innovation 81–2, 82–100
    'Special Area Plans' 89
    Urban Master Plans 89
'Special Area Plans' in Spain 89
SPIs *see* Social Performance Indicators
    (SPIs)
state
    globalization 71–2

governance 66–70, 70–2
Strategic Neighbourhood Approach 84
'surplus' concept in community economies 34
sustainability and microfinance 50

'*Temps de Cerises*' 13
territorial development
    social innovation 15, 16–21
    social relations 19–21
Territorial Innovation Model (TIM) 15
The New Deal for Communities 84
TIM *see* Territorial Innovation Model
'topographical proximity' 119
Trinitat Nova initiative in Barcelona 92–5
Turkish population of Berlin 104–7

Ueberfremdung (alienation) 109
United States
    'black success' 103
    deprived neighbourhoods 82–3
    ethnic entrepreneurship 102–4
United States Agency for International Development (USAID) 42–3
urban areas disintegration 16–17
Urban Master Plans in Spain 89

Urban Pilot Programme, European Union 85
urban policies
    European Union (EU) 101
    festivalization 110–11
    IAD 86–7
    marketization 110–11
urban territory in Latin America 120–7
USAID *see* United States Agency for International Development

Vick, Graham 157–8, 160
Vidigal slum (Rio de Janeiro) 124–5
Vietnamese population of Berlin 105–8

Weber, Max 12–13
welfare via top-down policies in Austria 132–6
World Bank 64
World Social Forum 25
World Trade Organization (WTO) 64, 73
Wozzeck (opera) 156–62, 163–4
WTO *see* World Trade Organization

Zentrum fur Soziale Innovation in Austria 139